IRINA RODICA RABEJA

WRITINGS III

ROBOTS & ARTIFICIAL INTELLIGENCE   TIME & STANDARDS

THERMOREGULATORS

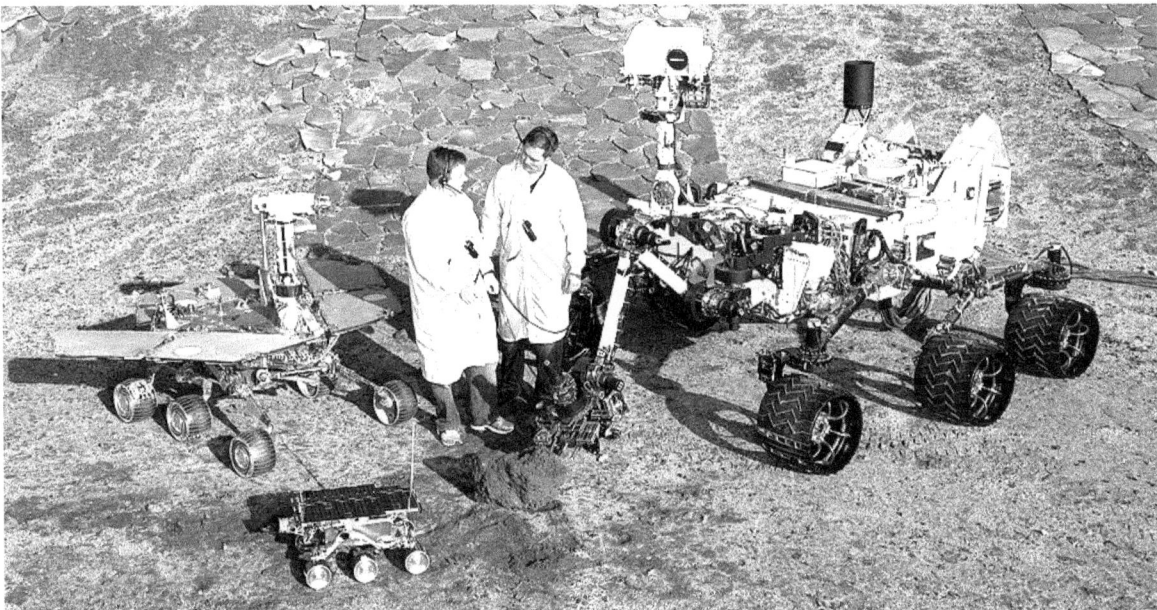

ENGINEERING   SCIENCE

IRINA RODICA RABEJA

WRITINGS III

ROBOTS &ARTIFICIAL INTELLIGENCE   TIME & STANDARDS

THERMOREGULATORS

ENGINEERING   SCIENCE

NATIONAL LIBRARY OF AUSTRALIA

A catalogue record for this book is available from the National Library of Australia

ISBN: 978-0-6486752-7-3

Publisher Irina Rabeja
Sydney Australia
2025

CONTENT

	Pg
ROBOTS / TYPES OF ROBOTS	9
TIME & STANDARDS	53
THERMOREGULATOR WITH THERMOCOUPLE	81
THERMOREGULATOR FOR THE FILAMENTS	99
IN VACUUM	
ANNEX	109

ROBOTS
TYPES OF ROBOTS

ROBOTS

PRESENTATION

Robots are machines able to act by themselves, they are automatically operated machines having internal/external control.

The word *robot* comes from Czech word *robota* meaning *drudgery* in English language. The appearance of the word *robot* is attributed to Czech painter and writer Josef Capek.

The robot is called also **bot**.

The robots replace humans in dangerous work places or perform repetitive, tiring tasks.

Mimicking humans or animals, the robots or bots exhibit intelligent behaviour.

The robots are able, by program - a discrete collection of instructions - to perform human-type actions and functions, without necessarily having a human appearance.

The robot with a partial or complete human-like appearance or one that performs human-like actions is called a *humanoid*.

The robot built to look like humans and perform some of the actions carried out by humans is called *android*.

If a human is being modified by addition of artificial limbs or organs, the human becomes a *cybernetic organism* or *cyborg* – 'cyber' is a prefix that generally refers to something related to computers, computer networks, internet.

All robots have the ability to move, some are fully mobile, able to dive in water, swim like fish, fly through air or move over ground with wheels or insect-like legs or caterpillar tracks.

The robots can be either relatively simple machines or incredibly complex, designed with a few core parts as the controller (computer program), the power supply and the mechanical parts including actuators, motors, drives, gearheads, bearings, links, joints translating signals in effects.

Some robots are designed for entertainment, others are designed for research in laboratories, but millions of them work every day as factories robots replacing human workers in dull, dirty or dangerous jobs.

All they require is the necessary maintenance.

The robots can operate in fierce heat or in unpleasant conditions such as sewers. They are suitable for repetitive jobs, such as packing objects or picking and placing items on a conveyor belt.

Thousands of industrial robots work in car factories where they spray vehicles with paint, weld or fuse parts together or help assemble vehicles.

Driverless mobile robots - automated guided vehicles AGV - move materials from one part of a factory to another, programmed to follow a bright painted line on the floor or to detect an electric signal from a buried guiding wire.

Some robots are tele-operated, controlled from distance by a human operator however perform by themselves some parts of their tasks.

Robots learn about their surroundings using sensors such as video cameras and proximity sensors. The information from sensors is analysed by the robot's controller, which makes decisions and then sends instructions to the different parts of the robot so that it can react correspondingly.

Robots' constituents are at the heart of the robot's ability to perform precise, accurate, repeatable movements. All parts are designed to perform with sensors that stimulate its dynamic performance.

The sensors enable robots to perceive and interact with their environment, providing them with valuable data to navigate surroundings and to engage with humans. There are a variety of sensors, categorized based on their primary sensing mechanisms:

- camera sensors
- colour sensors
- compass sensors
- force/torque sensors
- gas sensors
- hall effect sensors
- humidity sensors
- infrared sensors
- lidar sensors

- light sensor
- pressure sensors
- proximity sensors
- sound sensors
- strain sensors
- tactile sensors
- temperature sensors
- touch sensors
- ultrasonic sensors

A simple example of a robot is a machine that demonstrates the ability to manipulate objects around it by physically grasping them. Manipulation seems to be a prerequisite for the classification of robots: manipulators + systems that control their actions

The simple block schema of a robot is given below.

| SENSOR | → | CONTROLLER | → | MECHANICAL ACTIONS |

BLOCK SCHEMA OF ROBOT

The robot system decides the type of software suitable for its instructing or programming.

The program gives animation to the machine and it is called sometimes intelligence or artificial intelligence.

Robot Institute of America, a loosely organized federation of robot makers, defined the robot as:

"*A reprogrammable, multifunctional* manipulator designed to move material parts, tools or specialized devices through variable programmable motions, for the performance of a variety of tasks."

By *reprogrammable* they mean that if the robot gets a new assignment, it will need new instructions, but its basic structure will not change – ideally.

By *multifunctional* they mean that the robot is the mechanical counterpart of a general-purpose digital computer in that it can tackle various problems with no major hardware modification. Ideally the only thing that changes when the robot is reassigned, is its program of instructions.

The best of today robots are essentially teleoperators.

A teleoperator is a manipulator device, usually equipped with swivelling joints, controlled and watched over a distance by a human being.

Teleoperators were developed shortly after the World War II for handling dangerous radioactive materials.

In a modern robot the human operator is replaced by one or more microprocessors.

Current research efforts focus on creating a "smart" robot that can see, hear, touch and make decisions.

HISTORY

The idea of self-operating machines originates in the past, in the culture of ancient civilizations and developed along the centuries worldwide.

If you consider a robot as a general description for a mechanical, working replica of a living creature, man or animal, the robots existed 2000 years ago and in 18th and 19th centuries they were produced fairly in large quantities.

They were called then and are known today as *automaton* at singular, *automata* at plural. They were small moving mechanical figures developed from intricate spring, lever, gear and pulley-powered mechanisms used to make clocks.

Around 1495 year, Leonardo di ser Piero da Vinci (1452-1519) Italian polymath of High Renaissance, active as a painter, draughtsman, engineer, scientist, theorist, sculptor, architect sketched out - possibly made - a mechanical knight, which with cables and pulleys was designed to move in human-like ways.

The Model of Leonardo's robot with inner workings, possibly constructed by Leonardo da Vinci around 1495 is shown in the image below.

MODEL OF LEONARDO'S ROBOT

From ancient wooden puppets to modern computer characters in online games, people enjoined copying themselves. Humans are very complicated creatures with complex movement, brain and nervous system and powerful senses. Creating the human abilities and senses in a robot is a big challenge and scientists pursue it.

Robots or bots carry out tasks by themselves, yet it is considered that more interesting robots will still appear.

The idea of using robots to perform repetitious tasks quickly, cheaply and efficiently has intrigued human beings at least from the time of the Industrial Revolution in the eighteenth century. But not until the mid-twentieth century was the idea translated into reality.

The first design of a robot-type machine for industrial use in US is generally credited to *Plant Corp of Lansing* Michigan which in the year 1955 developed the PLANOBOT, a pick and place unit.

The next years followed UNIMATE and VERSATRAN.

In the year 1961, *General Motors* became the first major industrial firm to use robots. By 1970 year it expanded the use of robots to welding and automobile industry became the predominant user of robots in US.

By mid of years 1980, Japan with 250 robot manufacturers, leads the world in robotics development, production and application.

The first computer-controlled industrial robot T3 was developed in year 1973 by *Cincinnati Milacron*, forerunner of today intelligent robots, capable of using vision systems, sensors and responding to feedback.

In the 21st century, the robot is seen as a very powerful computer with equally powerful software housed in a mobile body and able to act rationally on its perception of the world around it.

There are famous robots:

LEONARDO DA VINCI

-The SKYWASH, a gigantic 33m long robot arm used in the year 1994 by Germany's airline *Lufthansa* to clean.

- KUKA TITAN, the world strongest industrial robot used in Germany in the year 2007. Its 32 m arm is able to lift 1000 Kg = 1 tone.

- WAKAMARU, Japanese robot made in the year 2005 by *Mitsubishi*. It is a 1m tall robot secretary on wheels.

- CYNTHIA is a 2m high robot bartender, built in the year 1999 by designer *Dick Becker* to pick up drink bottles and to mix 75 different cocktails drinks.

-DEEP BLUE II an *IBM* super-computer for chess playing, which beat the chess world champion Garry Kasparov in the year 1996.

- SOJOURNER was the first roving robot on Mars - 4 July 1997, 36 m/hour.

- Space builders built the International Space Station ISS between 1998÷2011. Crucial, integral role in its assembly, alongside human astronauts and mission controllers, played the robots: CANADARM2, ROBONAUT2, ASTROBEE Free space agencies including *NASA, Roscosmos, ESA, JAXA, CSA* have contributed to the ISS station's assembly.

- DA VINCI SURGICAL SYSTEM, fitted with 4 arms controlled by a surgeon sitting at a console. It is a robotic surgical platform built in the year 2000.
It was conceived initially by Leonardo di ser Piero da Vinci.

- ROBODOC, robot that drills a very accurate hole down a leg bone to fit an artificial hip implant, first surgical robot to operate on human in US, year 1992.

- ASIMO first humanoid to walk freely up, down stairs and around the corners. It was built in 2000 year. It stands for *Advanced Step in Innovative Mobility.*

- HUBBLE SPACE TELESCOPE is an unmanned robotic telescope, that has changed the face of astronomy and our understanding of the universe far beyond what was envisaged by Galileo Galilei (1564-1642), Italian astronomer, physicist, engineer, polymath, father of observational astronomy.
It is a large space-based observatory, launched in the year 1990, that orbits the Earth at 600km above the planet.

It observes the universe in visible, ultraviolet and infrared light, providing high resolution images and detailed spectroscopic data.

Hubble is joint project between *National Aeronautics and Space Administration NASA* and *European Space Agency ESA*.

It was initially designed for servicing by astronauts, now it operates remotely, controlled from the ground by NASA's Goddard Space Flight Centre.

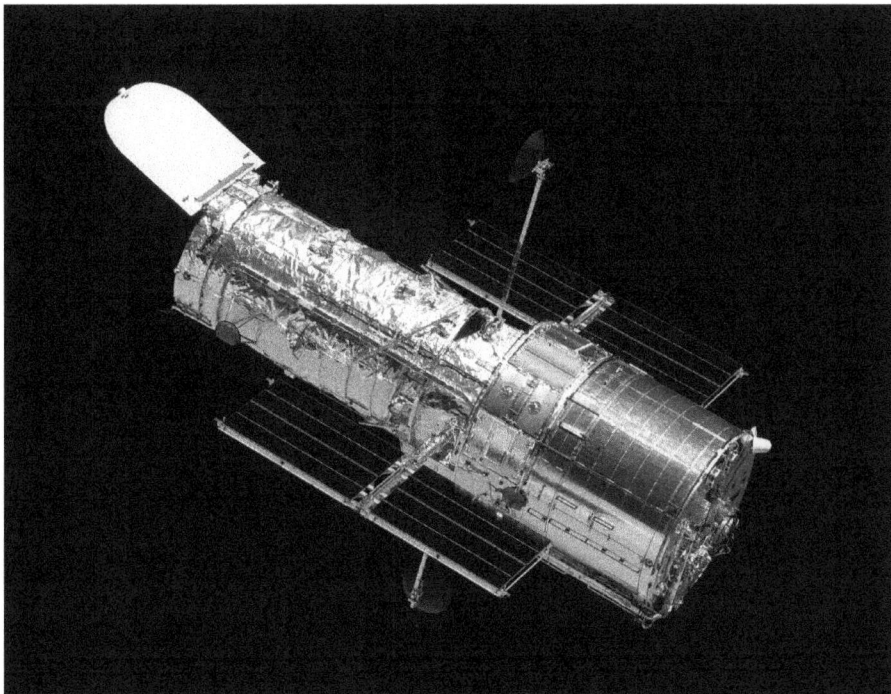

HUBBLE SPACE TELESCOPE

Along the time, the lifestyle of inhabitants on planet Earth encountered more transition periods known as Ages:

- Agricultural Age, also known as Agricultural/Neolithic Revolution.
It started around 10.000 BCE and it was a period of transition from hunter-gatherer lifestyle to settled agriculture.
It involved the domestication of plants and animals, food surpluses, population growth, development of villages and eventually civilizations.
In a subclassification there are two agricultural revolutions: first and second.

- Industrial Age, known also as Industrial Revolution.
It was the transitional period of the global economy toward manufacturing processes, succeeding the second agricultural revolution.
Manly was characterized by significant technological advancements, changes in economic practices and transformations in society.
It began in the year 1760 in Great Britain and later in other countries by replacement of hand tools with power-driven machines (power loom, steam engine) and concentration of industry in large establishments.
In a subclassification there are four industrial revolutions:
First Industrial Revolution began in late 18th century and continued in 19th century, years 1760-1840. Manly was characterized by Mechanical Production.
Second Industrial Revolution 1870-1914 was a phase of rapid scientific discovery, standardisation, mass production and industrialisation based on electrical energy.
Third Industrial Revolution began in mid-20th century, when nuclear energy and electronics entered the industrial landscape. It is known as Digital Revolution/Age based on Computer and Internet.
Fourth Industrial Revolution or The Smart Factory is based on artificial intelligence AI and information technology IT. It is towards automation and data exchange in manufacturing technologies and processes which include cyber-physical systems CPS, internet of things IoT, cloud computing, cognitive computing and artificial intelligence AI.

- Information Age, it started in mid 20th century.
Was a rapid shift from traditional industries, established during the industrial revolution, to an economy centred on the information technology IT.
It is the modern time regarded as a time when information is a commodity quickly, widely disseminated and easily available by computer technology.

- Intelligence Age
Now we are moving deeper in 21st century, away from Information Age, into Age of Intelligence. That means that understanding, analysing and processing data is becoming increasingly important.

Literature comments:

The last trillion-dollar industry was built on a code of 0s and 1s.
The next will be built on our genetic code.

The world has left the Cold War behind only to enter the Code War.
Industry of the future is the weaponization of code – a cyber-industrial complex.

ROBOTICS

Robotics is the technology branch which deals with the application, the design, the construction, the operation and the control of the robots. Robotics is a branch of automation, focusing on physical robots.

The word *robotics* was coined by the science fiction writer Isaac Asimov.

International Federation of Robotics IFR, established in the year 1987, promotes the robotics worldwide. According to IFR, in the 2022 year were active 3,903,633 industrial robots worldwide.

In the year 2025 IFR has identified the top 5 Robotics Trends as follows:

o artificial intelligence (physical, analytical and generative)

o humanoid robots

o sustainability through energy efficiency

o robotics as a business sector

o solutions to labour shortages

The robots appeared in science fiction, now robotics develops original solutions, building an efficient industrial automation.

No longer a luxury, the robotics is spread in industrial automation environment, it is an obvious trend in technology & innovation and fundamental useful to industrial development.

Robotics research has advanced in the last two decades by collaborating with other research units.

For aspiring engineers, a career in robotics can be very rewarding.

Due to the daily introduction of newer, quicker and more intelligent robots, the field provides fascinating professions for robotics engineers.

The top 10 Skills for Robotics Engineers are:

    o robotics fundamentals

    o mechatronics

    o programming

- knowledge of control systems
- machine learning
- sensors and actuators
- kinematics
- computer vision
- human-robot interaction
- strong problem-solving skills

The future is open to robots. Many future applications of robotics are foreseen although there are not yet means to build/deploy them.

Soft Robotics is the „Next Big Thing" in robotics.

To define it, it is to say that soft robotics takes advantage of materials that *bend, squish or stretch* or soft robotics may utilize material properties as *flexibility* (how much something can bend without breaking), *compressive elasticity* (how much something can be squashed and then return to its original shape) or *tensile elasticity* (how much it can stretch and then return to its original shape). These specific elements of softness are the most applicable to robotics because they allow for the possibility of repeated cycling.

The distinguish feature of soft robotics is the property called *compliance.*

In dictionary *compliance* word refers to conforming to a rule, such as law, regulation, policy or standard.

But *compliance* is also a technical word for softness of objects/systems when they are touched and by referring to how they act, *embodied compliance* describes objects/systems' reaction when encounter a force: instead to snap they squish, stretch, bend, compress or expand

Other kinds of tactile softness like smoothness, silkiness, plushness, are suitable for roboticists to utilize too, either in how those properties relate to human interaction or for their physical consequences for things like friction.

AUTOMATION

Automation is defined by the technologies that remove human intervention in work, achieved by hydraulic, mechanical, pneumatic, electric/electronic devices controlled by computers.

The branch of automation, focusing on physical robots is the robotics.

The advancements in robotics and automation imply computing and they all are changing the way how people work and live.

The word *automation* derived from Greek word *automatos* meaning acting on oneself, one's own will.

Automation Made Easy is a reality that is reshaping the manufacturing landscape. It is a move to more efficient, smart operations where technology takes on hardship, allowing human creativity and strategy to be useful.

It underlines a transformative era in manufacturing, better efficiency and innovation. It orders businesses' operations, with better productivity and advanced automation technologies.

As businesses navigate the 21st century, automation emerges as a guiding of innovation.

At *Automate 2024*, North America's largest robotics and automation event in May 6-9 2024 with over 800 exhibitors, the show floor hosted leading automation solutions from all world, from robotics to vision to motion control, artificial intelligence AI and more.

It showed that the automation is no longer in the future, it is here and now with a clear message: embrace automation, simplify your operations to pave the way for future success.

The journey toward seamless automation begins when new technologies blend with existing systems effortlessly.

Automated processes reduce the potential for human error, provide better production times and ensure quality.

Automated solutions are designed with user-friendly interfaces, ensuring that they can be operated by many users.

Automating hazardous tasks significantly improves worker safety.

At present 5 connectivity megatrends are shaping industrial automation, 5 industrial trends are driving innovation and setting new standards for efficiency and productivity:

o Industrial Internet of Things IIoT Adoption

Sensors and Internet of Things IoT gateways connect the data to the cloud using Bluetooth or Long-Range Wide Area Network LoRaWAN to transform raw data into actionable insights.

From off-the-shelf to custom solutions, IoT adoption in industrial settings is essential for next-gen applications.

o Smart Manufacturing

It is within factories where machines communicate effortlessly, making instantaneous decisions that boost productivity.

o Predictive Maintenance

Wireless connectivity for IoT sensors and gateways, allows continuous monitoring of equipment health.

o Robotics & Machine Learning

Smart manufacturing and need for higher efficiency have driven the adoption of robotics. Modern factories and warehouses now rely on autonomous robots to perform complex tasks and optimize operations.

o Next-Gen Warehousing

e-commerce has dramatically increased the demand for warehouses.

They are vastly different today, as *sensors* track inventory in real-time, *robots* handle stock movement and *advanced analytics* optimize storage layouts.

ARTIFICIAL INTELLIGENCE

Artificial Intelligence concept appeared in 1956 year, when John McCarthy, professor at Dartmouth College, organized a summer workshop to clarify and develop ideas about *thinking machine*. He chose the name Artificial Intelligence for the project.

AI is acronym and stands for artificial intelligence.

The terms artificial intelligence and robotics are freely used and interchanged. The combination of artificial intelligence and robotics, the convergence of artificial intelligence algorithms and advanced robotic hardware, creates intelligent machines that can perceive, learn, adapt with precision & flexibility. However, the physical nature of a robotic system is different from the pure abstraction of artificial intelligence.

In the world of robotics, collaboration between professionals from different fields of human activity is successful by combining the issues of cognition (perception, awareness and mental models) and the physical attributes (safety, dependability and dexterity)

Much attention is given to the social and medical benefits of robots.

We are experiencing a transition from **Information and Communication Technology ICT** to **InterAction Technology IAT**.

Wireless technology through communication (Bluetooth, Wi-Fi, 5G) and power transfer (wireless charging) will help new generations of robots, free to roam controlled via wireless links, using computing and data storage of the cloud.

The relationship between hardware and software design opens new possibilities, from manufacturing processes to space exploring, with the characteristics:

- o Enhancing Robotic Precision and Accuracy
- o Adaptability and Autonomous Decision-Making
- o Better Workflows and Human-Robot Collaboration
- o Boost labour productivity 40% and automate 38% jobs

The artificial intelligence has been explored on many research frontiers, where the most cutting edge - the latest and most advanced stage - discoveries and developments are happening.

The elusive goal of the artificial intelligence is the creation of a machine thinking indistinguishable from a human.

Some of the first insights in the matter are:

– Alfred North Whitehand (1861-1947) English mathematician and philosopher best known for the book Principia Mathematica and the creation of the philosophical school known as Process Philosophy said:

"Civilisation advances by extending the number of important operations we can perform without of thinking about them."

– Alan Turing (1912-1954) English mathematician, computer scientist, logician, cryptanalyst, philosopher, theoretical biologist, considered the father of modern computer science, in the year 1950 suggested that:

"If the human believes the computer is another human than the computer exhibited artificial intelligence. Robotics and artificial intelligence are fundamentally attempts to model various aspects of ourselves. "

– Marvin Lee Minsky (1927-2016) American cognitive scientist and computer researcher, concerned largely with research in artificial intelligence wrote extensively about artificial intelligence and philosophy and in the year 1960 defined:

"Artificial intelligence is the science of making machines do things that would require intelligence if done by men.

Intelligence is some mental processes enable us to solve problems we consider difficult.

Intelligence is our name for which-ever of those processes we don't get understand."

– Stephen William Hawking (1942-2018) English theoretical physicist, cosmologist, author, director of research at Centre for Theoretical cosmology at University of Cambridge said:

"The rise of powerful artificial intelligence will be either the best or the worst thing ever happened to humanity. We don't know which."

The concept of artificial intelligence appeared in 1956 year, two years after the death of Alan Turing, when John McCarthy, professor at Dartmouth College, New Hampshire organized a summer workshop to clarify/develop ideas about *thinking machine* and chose the name Artificial Intelligence for the project. Artificial Intelligence emerged nowadays because there are Large Datasets from Internet and digital technologies.

Especially in the machine learning ML and the deep machine learning DML, having access to vast amounts of data is crucial for training models successfully. The artificial intelligence AI can contain machine learning or not, for example: Human Languages Robots have programs called Large Language Models LLM using specifically deep learning models, trained on massive amounts of text data to understand and generate human language – that is artificial intelligence with machine learning.

Chess Playing Robots have programs using a set of instructions based on the technique called alpha-beta pruning – that is artificial intelligence without machine learning.

Artificial Intelligence is a broader term for virtually any software used in programming of robots performing tasks that typically require human intelligence.

In other formulations, artificial intelligence is a set of technologies that enable computers to perform a variety of advanced functions, including the ability to see, understand and translate spoken and written language, analyse data, make recommendations and so on.

ALAN TURING

Until now the artificial intelligence has outperformed humans on a variety
of language understanding and visual understanding benchmarks.

But the foundation models of artificial intelligence still lack advanced reasoning
and planning capabilities.

Advanced reasoning and planning capabilities for artificial intelligence are now
in development, with companies like OpenAI Inc. and Meta Platforms Inc.
giving results integrated into products like Google Search.

However, *achieving human-like reasoning and long-term planning remains
a challenge for research.*

Research beyond pattern matching to genuine problem solving and strategic
thinking and replicating human-like reasoning is still daring.

Besides, advanced reasoning models imply intensive computing and have
higher latency compared to traditional models.

It is remarkable that the concept of artificial intelligence can be found in the ancient literature.
Around the 8th century BCE, the Greek poet Homer wrote in his epic poem 'Iliad' about *Hephaestos*,
the god of artisans, blacksmiths, carpenters, craftsmen, fire, metallurgy, metalworking, sculpture and
volcanoes who built golden automata, self-operating machines to help him work.
Hephaestos was an able inventor, he built ATTENDANTS for himself with
"intelligence in their hearts" and looking as young women, he built AUTONOMOUS VEHICLES
that could travel to and from the home of gods and a lethal autonomous weapon system named
TALOS that patrolled the beaches of the island Crete.

See further:
The sculpture Vulcan/Hephaestus, Marble, Louvre, reception piece for the French Royal Academy 1742 by
artist Guillaume Coustou the Younger (1716-1777)

HEPHAESTUS AT THE FORGE

TYPES OF ROBOTS

The robots can be classified function of their design, configuration, functionality, level of autonomy and so on.

Further it is given an enumeration of robots, based mainly on their area of application.

INDUSTRIAL ROBOTS

The industrial robot, is an automatic, programmable, transfer and handling machine. Industrial robots have begun to revolutionize the industry. These robots do not look or behave like humans but they do work like humans. Industrial robots are advanced automation systems that utilize computers as an integral part of their control. Computers are a vital part of industrial automation. Drives for computerized robot systems are categorized by their power source into hydraulic, pneumatic and electric systems, with electric drives being the most common. These drives, in conjunction with actuators and transmission systems, are essential for controlling robot's motion, precision, accuracy and efficiency.

Robots can be designed so that they accomplish many different tasks.

The very important industrial robots are used industrywide for pick and place, sorting, packaging, material handling, palletizing, labelling, assembly, welding, gluing, spray painting, deburring, drilling, die casting, loading machine tools, mining, inspection etc.

Palettizing Bread Factory Robots

KUKA Industrial Robots manufacturer offers a comprehensive range of industrial robots helping diverse applications with precision, flexibility and efficiency.

In the image below there are robots used in food production, for palletizing at a bakery.

PALLETIZING ROBOTS

Autonomous Mobile Robots

Autonomous Mobile Robots AMR are robotic vehicles that can handle, move and navigate materials independently without human intervention.

They are robotic systems designed to operate and move within a defined space, autonomously making decisions based on their environment and tasks.

They can safely work everywhere in factories, warehouses, homes without humans' supervision. They are typically designed with sensors to avoid obstacles, while performing their designated task and with their independence and agility they are changing the logistics of the warehouse.

At Fair Center Munich, **Automatica 2025** *on 24-27 June 2025, leading exhibition for Smart Automation and Robotics, introduced* Autonomous Versatile Robotics *–the* vision *for the next generation of industrial automation.*

Enabled by generative AI, the creative AI, that is where mobile robots can plan, adapt and switch between tasks in real time without human intervention, built on six core capabilities: human interaction, sensing and perception, autonomous reasoning, motion control and safety, navigation, dexterity

"We've given robots eyes, hands and mobility," said Marc Segura, President of ABB Robotics, *"Now, with generative AI, we're building their brains – so they can understand, adapt and act on their own."*

Automated Smooth Sorting Robots

Automation, material handling & distribution sorting systems are very important.

SEW-EURODRIVE company's automated sorting solutions offer increase capacity, reliability performance and efficient energy usage.

Tompkins Robotics automation company, a provider in robotic sortation solutions for distribution and fulfilment operations, launched the latest autonomous mobile robot for sortation: tSortPost

tSortPost offers benefits to parcel sortation, including higher speed, customizable options with a cross-belt or tilt tray divert mechanism, an adjustable elevation up to 1.5 meters and minimal infrastructure requirements without extensive hardware.

tSortPost is driven by the *Transcend* software suite, optimizing performance, seamless integrating with the systems and providing versatile and efficient solutions.

Human-centric Robot fleets for performing repetitive tasks

In the process of executing the efficient transportation and storage of goods, the partnership between *Ricoh North America* and *Agility Robotics* companies is a collaboration to support automated warehouse solutions and managing humanoid robot fleets.

Ricoh North America company is extending its Service Advantage program, a lifecycle management solution.

Agility Robotics company is the creator of humanoid robot DIGIT, a leading bipedal Mobile Manipulation Robot MMR. a human-centric robot capable of performing repetitive tasks.

Ricoh North America will support DIGIT robots and *Agility Arc*, Agility's cloud automation platform for operational management and troubleshooting, to deploy & manage DIGIT fleets.

Gluing Robots

Well-known as automated applications within the manufacturing sector are material handling, assembly or pick and place tasks.

Less well-known applications as gluing and dispensing (process of applying adhesives) are becoming more used across various industries.

Adhesives are essential in manufacturing eliminating the need for other forms of bonding like screws or nails.

Because of necessary accuracy and safety concerns, automated gluing is superior to manual gluing.

The possibilities of gluing robots and how you can utilize automated gluing to reduce costs considerably while improving safety, reliability and efficiency are considered and embraced by companies.

Applications that prove gluing robots as necessary are:

- o Furniture gluing
- o Window gluing
- o Mirrors gluing
- o Battery console gluing
- o Liquid gasket
- o Air filter gluing

Vision-Guided Robots VGRs

Vision-Guided Robots VGR, has reshaped what robots are capable, perceiving, understanding and interacting with surroundings like never before.

KINE Robotics company, a provider of robotics solutions in Finland, together with *Basler* company's distributor *OEM Finland*, has developed a conveyor tracking system that enables a Pick-and-Place ROBOT with 3D COMPUTER VISION to safely grip bags of bread loaves and rolls of different sizes and shapes on the conveyor belt and to place them in transport boxes.

The previously manual picking of baked goods was thus replaced by a high-performance system with *3D Vision Guided Robotics,* faster, less error-prone.

3D Ultrasonic Sensor Robots

Autonomous Mobile Robot AMR emerged worldwide due to efficacity in material handling and labor shortages in manufacturing, logistics and warehouse.

AMRs are versatile robots which move safely around objects and people to the level described in the EN/ISO 13849/SIL2 machine safety standard. Today a typical AMR sensor includes safety-certified 2D LIDARs and depth cameras, an expensive and computationally intensive combination. The next generation vision sensor for autonomous fleets has Sonair, a 3D Ultrasonic Sensor, enables to provide safe, fast and effective obstacle detection, seeing objects that laser/camera struggle to see, consuming less energy with lower computational demands.

Tactile Robots

Tactile robots are robots which embody tactile intelligence.

The tactile sensing gives to robots the comprehensive understanding of their surroundings and there are many applications where tactile sensing flourish in robotics: delicate manipulation, precise grasping, complicate object recognition, seamless human-robot interaction.

Tactile sensing has a ground breaking role in reshaping robotics and interactive intelligent systems.

Tactile sensors, built to reproduce the complex sensitivity of human touch, empower robots with the ability to perceive and interpret the multifaceted environment, enrich robots with a deeper understanding of their surroundings and enhanced their interaction capabilities.

Robotic Arm with Detachable Hand

In manufacturing plants, the arm-robots are large and heavy, attached to floors or tables, that limit their reach.

At *École Polytechnique Fédérale de Lausanne* EPFL, at robotics lab of EPFL's *Learning Algorithms and Systems Laboratory* LASA was designed a robotic arm that can detach its hand to reach and grasp objects.

The robotic hand can move unaided to grasp different objects and carry them back to the arm.

To develop it, the researchers generated and refined a basic design via a genetic algorithm based on biological evolution and the MuJoCo physics simulator. MuJoCo is a free and open-source physics engine that aims to facilitate research and development in robotics, biomechanics, graphics and animation.

Metal Die Casting Robots

Factory automation facilitates a more efficient, cost-effective and safe manufacturing process. Once, automation involved controls that were mechanical but now automation frequently includes electronic and computer controls, based on high-level programming languages.

As example of automation in a factory is the industrial robot for metal die casting in the foundry industry seen in the image further.

It defines also the robotics in metal manufacturing.

METAL DIE CASTING ROBOT

COLLECTIVELY PROGRAMMED ROBOTS

Collective robotics is an emerging and promising research field, where more robots work together as a team, group or swarm to accomplish a task.

Robot Swarms

Swarm robotics exhibits the strength of the collective; it is the arrangement for a large number of mostly simple physical robots to work together to perform a task efficaciously.

Robot swarms could be used to monitor radioactive locations in underwater environments and in critical search and rescue missions in areas hit by natural or human-made disasters.

COLLABORATIVE ROBOTS – COBOTS

In the year 2023 Tara van Geons wrote in her article

"Emergent Robotics: Pioneering Solutions in Modern Industrial Automation"

that the **cobot** is a pioneering solution.

Collaborative robot or **cobot,** is a robot that can work close to a human being.

The collaborative robots, cobots, are changing the industrial landscape.

They enable work environments where humans and robots complement one another's capabilities. To avoid accidents, they have built-in sensors able to see/feel the humans, so they are safe.

They are a different aspect of automation, from the usual industrial robot.

Collaborative robots can be used in different areas of manufacturing, for ex. pick and place, packaging, screwdriving, soldering, and mounting components.

Any company of any size, can access the cobots technology, helping to fill the productivity gap, to boost efficiency.

Besides being highly adaptive and more accessible to manufactures of all sizes, cobots are designed to be less complex and are more affordable.

The payback period of a cobot could be measured in days.

Yaskawa HC series Cobots

Yaskawa HC series cobots are highly versatile, portable human collaborative robots and are ideal for variety tasks as:

- o Machine tending (automation of repetitive tasks)
- o Material handling
- o Packaging & palletising
- o Welding
- o Light assembly

HEALTHCARE ROBOTS

Healthcare robotics works together with electrically powered medical devices, with the aim of improving patient care.

Motion control technologies and techniques are crucial for healthcare robotics engineering applications, typically having very rigorous functional requirements in areas such as safety, reliability, tolerances, cleanability, sterilization with key challenges as non-cartesian motion management, dynamic load management and others.

High Performance Robotic Joints HPJ

The high-performance robotic joints HPJ, are ideal for surgical robots, cobots and manufacturing applications. The individual components of a joint should work together seamlessly and with high precision. To provide the best possible product for robotic joint systems, *maxon* company has designed and manufactured a high-performance robotic joints modules line.

It created the robotic joints module called HPJ57, a compact, low-profile robotic joints module configured for lightweight use.

The module integrated devices are:

- an electronically communicated EC and frameless Dynamics Torque DT motor with fine, smooth and predictable motion, at 48 VDC supply and output speeds in the range 15-65 rpm.

- a strain wave gear which can achieve much higher reduction ratios of up to 30 times

- a motor encoder to sense the motor's speed and position and send feedback signals to control components in applications

- an output absolute encoder that directly output the exact position of the shaft it is measuring, one of the more critical components of the robotic joint modules, responsible for the precision of the joint itself

- an optional braking device however essential if it is to hold position when power is lost
- a high-accuracy position controller, through a ready-to-connect, high-performance, small, completely digital, intelligent positioning controller EPOS4

Breakthrough motion control / Performance motion devices

A breakthrough motion control drive from *Performance Motion Devices Inc.* company in year 2024, is a smaller controller that results in decreased design time, improved performance, lower cost.

ION®/CME N-Series Digital Drives are compact, PCB-mountable modules that provide high performance motion control, network connectivity and power amplification for Brushless DC, DC Brush and step motors.

N-Series are the only drives in the market that deliver up to 1KW of power and come in a tiny footprint (37.6mm x 37.6mm x 16.8mm) with an enclosed package that protects the electronics inside.

Precision drive solutions in surgical Robotics

Advanced precision drive solutions in minimally invasive surgeries MIS provide precise, rapid, fine controlled movements.

Those solutions enable for surgical robots to perform intricate procedures with highest accuracy, steadiness and reliability. These are high-tech machines with fine-tuned integration of various components working flawlessly.

A Precision Drive System in Surgical Robotics is made by;

Motors, Actuators, Precision bearings, Custom gear, Gear assemblies, Small mechanical components

Major Companies Developing Surgical Robots are:

Intuitive surgical, Medtronic, Ethicon, Stryker, Zimmer Biomet

Future of surgery: intelligent Robots

Surgical intervention is reliable, complications are reduced and recovery times are shortened in the new age of healthcare which is arriving.

For that are used the augmented intelligence and real-time imaging.

Augmented intelligence is a subset of artificial intelligence AI in which AI technologies assist humans rather than replace them.

Humans use it to enhance their capabilities and tools.

Real-time imaging is the rapid acquisition and manipulation of ultrasound information from a scanning probe by electronic circuits to enable images to be produced on TV screens almost instantaneously.

LUNA *of Asensus Surgical* is the next generation digital surgery platform, elevating the standards of robot assisted surgery:

- incorporates advanced hardware and software as a surgeon console, an interactive touchscreen and an Ultra-HD 3D monitor for enhanced visibility

- has four robotic arms for various procedures and an unique instrument drive for a range of advanced tools

It incorporates the digital powerhouse of the SENHANCE SURGICAL ROBOTIC SYSTEM, a new robotically-assisted surgical device RASD that can help facilitate minimally invasive surgery and the Intelligent Surgical Unit ISU that enriches the surgical experience with its haptic feedback - which provides surgeons the ability to "feel" the anatomy through the robot as if they were directly touching it - its eye-tracking camera control - that moves the camera based on the surgeon's gaze - and its 3D visualization for a comprehensive view of the surgical field.

The ISU's augmented intelligence capability enhances intraoperative procedures with 3D digital measurements of anatomical structures to millimetre precision, giving accurate straight-line and curved measurements.

Treatment of brain bleeds with Nanorobots

University of Edinburgh researchers found a solution to deal with the bleeds caused by aneurysms - the ballooning and weakened area in an artery.

They developed magnetic nanorobots, twentieth the size of human red blood cell, charged with blood-clothing drugs, enclosed in a coating designed to melt at certain temperature and guided using magnets and medical imaging to the aneurysm. There, heated at the melting point of the coating, the nanorobots let out the blood-clothing protein to avoid bleeding in brain.

ROBOTS DERIVING FROM NATURE & LIFE

The creators of robots have taken models from nature and life,
by bio-mimicking or by bio-inspiration.
Biomimetics replicates specific biological structures. Mimicking humans, animals, insects or plants, the robots exhibit intelligent behaviour.
Bioinspired draws inspiration from nature to develop new solutions.

Plant-based Robots

The plant-based robots can help with reforestation.

The researchers at the *Italian Institute of Technology* IIT in collaboration with the *University of Freiburg Germany* have developed a biohybrid robot made from flour and oats which moves in response to humidity and is being tested as a means for reforestation efforts.

The biohybrid robot replicates the wild oats of the genus *Avena,* very successful grasses which colonized the globe, due to the ingenious design of their seed-carrying fruit. The fruits are designed to jump and dig without a metabolism or muscles. Consisting of a small seed package and two bristle-like appendages called awns, the fruit is a mix of living and dead tissue.

The spindly, thin awns, are made of dead tissue and in humid or wet conditions, a section of cells at the base tilts, causing the awns on either side of the fruit to entangle, storing energy. When that is released, push the fruit forward into soil cracks where it embeds with the fine hairs along its side.

Bio-inspired Robots

An example for a robot inspired by life is the Bio-inspired BIG DOG quadruped robot from the illustration further, developed as a mule that can traverse difficult terrain.

BIG DOG ROBOT

AGRICULTURAL ROBOTS

Robots are used also in agriculture, to automate tasks, to increase quality and efficiency. They have artificial intelligence using machine learning, computer vision and perform functions as seeding, harvesting, weed control, spaying and livestock management.

Picking Precision Machine

In agriculture, the community can no longer rely on human harvesting of the food. In grain fields, machine harvesting has been achieved for decades, but harvesting fruits and vegetables has been a problem.

For example, the strawberries are one of the most labour-intensive crops and without an automated harvesting way, they will remain high-priced. *Harvest CROO* company is transforming agriculture with technological advancements, with increased efficiency and productivity, by developing a 12.5-ton connected, integrated, automated robotic strawberry harvester. The whole harvester is an Internet of Things IoT device on wheels.

"The project is continuously driven because of the difficulty of finding harvesting labour," said Bob Pitzer, co-founder, CTO and engineering manager at *Harvest CROO Robotics LLC*.

Laser Weeding Machine

A robot vision system is deployed on a machine which destroys the weeds. "Laser Weeding Machine" was created to detect and destroy the weeds on field. Paul Mikesell, CEO of *Carbon Robotics*, a US-based company that develops autonomous weeding robots for farmers, explains that the robot vision system on the LASERWEEDER machine works by training of AI models to differentiate the "ugly" from saleable crops.

SOFT ROBOTS

The emerging field of soft robotics provides opportunities for problem solving in challenging fields as those exposed to chemicals or radiation or in the emerging environment of space exploration.

Electronics-free Soft Robots

The application range of metallic machines can be realized by soft robots too. Due to their properties, soft robots could be used to perform tasks that cannot be carried out by conventional robots in challenging fields, where are exposed to chemicals or radiation that would harm robots made of metal or on terrain that is difficult to access.

That requires soft robots to be controllable without usual logic electronics which involves metal components.

A team of researchers from the *University of Freiburg*, led by Dr Stefan Conrad and Dr Falk Tauber, has developed 3D-printed pneumatic logic gate modules for functions AND, OR, NOT that can control the movements of soft robots using air pressure, making possible to produce flexible and electronics-free soft robots.

Potential benefits of expansion of soft technology in space exploration

Conquering the space is very expensive, so for ex. to get anything to low Earth orbit LEO is $1000-10000/pound. Soft materials have a good strength-to-weight ratio, which could be used for maximum advantage.

Besides a weight limit, rocket payloads have a volume limit. A compact profile that can expand at necessity is an advantage that softness can offer.

Gravitational forces and vibration can damage/destroy some hard components. But soft parts have tolerance for the gravitational force and vibrance.

NASA proposed a SOFT ROBOTIC ROVER for exploring the moons Europa (Jupiter) and Enceladus (Saturn)

Bioinspired Soft Robotic Rovers

Soft robotics grew from the idea of biomimetics which over time shifted to the idea of bioinspired.

SQUID-LIKE ROBOT, a variation of SOFT ROBOTIC ROVER, applies the mechanics of biology to exploration. This rover can travel where wheeled vehicles cannot and where solar power and nuclear power are not accessible. It scavenges electrodynamic power from locally changing magnetic fields. The moons Europa of Jupiter and Enceladus of Saturn, whose oceans lie beneath a thick layer of ice are two potential targets.

DRONES

The drones are uncrewed aircrafts or vessels guided by remote control or onboard computers. Here is an example of their use:

Drones are helping European data centres

The drones are being used to monitor data centre developments in Northern Europe through a partnership between *DroneDeploy* and *Globhe* companies. *DroneDeploy* is an aerial and ground reality capture platform.

Globhe is the largest drone data marketplace, making drone pilots available for customers. Through *Globhe*, over 11,000 local drone operators in 147 countries are being tasked with mapping and inspecting sites.

Many are using the *DroneDeploy* software for data capture. The organizations seamlessly integrate data from drones in their planning, monitoring and evaluation, because they can tap into drones worldwide through an interface.

MILITARY ROBOTS

There are a wide variety of applications for military robots, for example:

IED detonator Robot

There are robots built to detonate random bombs, a dangerous task for humans. In the illustration, an U.S. Marine Corps technician prepares to use a telerobot to deploy a device that will detonate a buried improvised explosive device IED near Camp Fallujah, Iraq, Nov. 27, 2005.

TELEROBOT USED TO DETONATE EXPLOSIVES

ROVER ROBOTS

A rover robot is a space exploration robotic vehicle used particularly in exploring the land of a planet.

The rover vehicles are used to explore rough terrain and got their name from *wanderer* or *rover*, someone who does not settle in one spot.

Rovers have several advantages over stationary landers:

they examine more territory, they can be directed to interesting features, they

can place themselves in sunny positions etc.

Three Mars Rover Robots

In the image below are presented two roboticists with three Mars rover robots.

- front and centre, the flight spare of SOJOURNER landed on Mars in 1997

- left, test vehicle Mars Exploration Rover MER, working sibling
  of SPIRIT and OPORTUNITY landed on Mars in 2004
- right, test rover for Mars Science Laboratory that landed CURIOSITY
  on Mars in 2012

*CURIOSITY is a car-sized Mars rover exploring Gale crater and Mount Sharp on Mars
as part of NASA's Mars Science Laboratory MSL mission.
Curiosity was launched from Cape Canaveral on November 26, 2011, at 15:02:00 UTC
& landed on Aeolis Palus inside Gale crater on Mars on August 6, 2012, 05:17:57 UTC.*

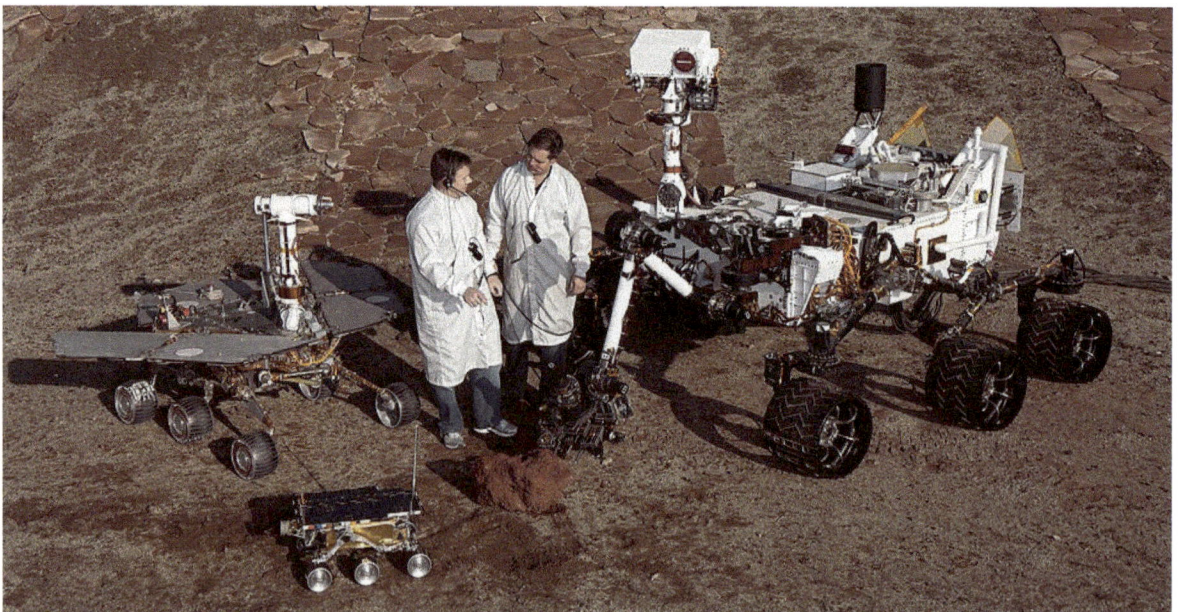

THREE MARS ROVER ROBOTS

references

Iliad, book by Homer ~ 8th century BC

Robots and Robotology, book by R.H. Warring 1983

Advanced Robot Systems, book by Mark J.Robillard 1984

The robot revolution, book by Tom Logsdon 1984

Dictionary of robotics, book by Harry Waldman 1985

Robotics for engineers, book by Yoram Koren 1985

Robotics in service, book by Josef F. Engelberger 1989

Robots from science fiction to technical revolution, book by Daniel Ichbiah with introduction by Will Wright 2005

Industrial robotics: selection, design and maintenance, book by Hary Colestock 2005

Robots, book by Clive Gilford with illustrations by Frank Picini 2008

The industries of the future, book by Ross Alec 2016

The Hubble Space Telescope, book by David J. Shayler 2016

Make: Soft Robotics, book by Matthew Borgatti & Kari Love 2018

The Reasonable Robot / Artificial Intelligence and Law, book by Ryan Abbott 2020

Drive Systems for Robotics & Automation – FAULHABER

Adaptive Computing in Robotics – AMD-XILINX 2022

Emergent Robotics: Pioneering Solutions in Modern Industrial Automation
article by Tara Van Geons / Nov 2023 MachineDesign INDUDSTRY MARKETS>ROBOTICS

"Sorting Smoothly | Sorters | SEW-EURODRIVE"
Video on YouTube Nov 2023

The AVNET white paper: How to design a cobot 2023

Announcing the Addition of tSortPost: A New Pedestal AMR for Parcel Sortation
article by Jim Serstad / July 2024 Eureka!

Robotics at IMTS 2024: Show Launches, Market Ripples, Partnerships and More
article by Rehana Begg / Sept 2024 MachineDesign INDUSTRY MARKETS>ROBOTICS

5 Applications that Prove Gluing Robots are the next „Big Thing"
article MachineDesign ROBOTICS&AUTOMATION May 2024
3D ROBOTICS IN THE BAKERY INDUSTRY
article #Machine Vision of OEM AUTOMATIC

New 3D Ultrasonic Sensor Technology Dramatically Reduces Costs and Improves
Safety for Mobile Robot Developers by Knut Sandven / MachineDesign INDUSTRY
MARKETS-ROBOTICS / Oct 2024

YASKAWA HC series Cobots
Robotic Automation Pty Ltd leading supplier of robotic systems

Key Component Integration for High-Performance Robotic Joints
article by Biren Patel / March 2024   MachineDesign INDUSTRY MARKETS-ROBOTICS

ION®/CME N-Series Digital Drive   PERFORMANCE MOTION DEVICES | MOTION CONTROL AT ITS CORE

University of Edinburgh researchers develop treatment of brain bleeds   *by Roshini Bains*
Eureka! Sept 2024

The important Role of Precision Drive Solutions in Surgical Robotics by Linda Shuett,
SDPI/SI Ideas in Motion / Feb 2025

Picking Precision: Advantages of Robotics in Agriculture
article by Sharon Spielman May 2024   MachineDesign   INDUSTRY MARKETS>ROBOTICS

Deep Learning Vision in Agriculture   article by Sharon Spielman   MACHINE DESIGN 2024

These plant-based robots will help with reforestation   article by Bradley van Paridon
May 2024 ADVANCED SCIENCE NEWS / Earth and environment
3D-printed digital pneumatic logic for the control of soft robotic actuators
article by Stefan Conrad, Joscha Teichmann, Phil Auth, N. Knorr, Kim Ulrich, D.
Bellin,Thomas Speck, F. Tauber
published in Journal *Engineering, Materials Science. Science Robotics*   Jan 2024
Drones are helping European data centre operators monitor new builds
article Sean Buckley July 2024   LIGHTWAVE + BTR

RoboBusiness Event in Santa Clara California   16-17 October 2024   Simon Brazier

Intelligent Robots are the Future of Surgery, article by Carolyn Mathas Talk Blog
of Mouser ELECTRONICS April 2024

Integrating AI into Robotics: The Fusion of Hardware and Software Design /Dev Nag
ElectronicDesign Markets.Automation Sept 2024

Detachable Hand Extends Robotic Arm's Reach, article by Cabe Atwell /
ElectronicDesign Markets.Automation Oct 2024

Robotics Meets AI & 5G — The Future is Now! by *Bruno Siciliano / University of Naples*
Federico II, Italy

Embodied Tactile Intelligence by Dr Mohsen Kaboli / RoboTac @ BMW Group
Content Strategy Learning / Linkedin
‚Automation Made Easy' from SEW-EURODRIVE April 2024

5 CONNECTIVITY MEGATRENDS SHAPING INDUSTRIAL AUTOMATION
Ezurio Aug 2024

AUTOMATE 2024 SNAPSHOT: Robotics, Automation, AI and More 2024
from Machine Design Library - A compendium of technical articles from *Machine Design*
and *maxon* 2024

Technology Predictions Report 2025 / IEEE Computer Society

Autonomous Mechanical Systems: Designing Smart Robots 2025
from Machine Design Library - A compendium of technical articles from *Machine Design*
and *maxon* 2025

Wikimedia   Wikipedia

TIME & STANDARDS

INTRODUCTION

From old days, the time in general and its measuring, display and recording in particular, have been problems that concerned the people from almost all fields of the human activity, being often a determinant factor in their work.

The devices for measuring and displaying the time have been called usually clocks. The clocks have been of different kinds, function of the talent or the ingenuity of the manufacturers, the selection being done function of their simplicity and precision.

In our times many clocks are electronic clocks with digital display.

By the help of electronics, the clocks precision increased very much because the measurement of time is done by the means of the temporal frequency standards, between time and frequency there are not fundamental differences.

Frequency is the number of occurrences of a repeating event per unit of time.

Frequency is an important parameter used in science and engineering to specify the rate of oscillatory and vibratory phenomena, such as mechanical vibrations, audio signals (sound), radio waves and light.

The time and the frequency are non-perceptible quantities, they are impossible to be perceived by our senses or mind.

The time and the frequency, are based on two aspects of the same phenomenon and they can be measured only in connection with a physical entity.

The time and frequency measurements are useful in very many fields of human activity, helping the scientists, specialists, researchers, engineers united in the common effort to master the phenomena which were revealed by Nature and to penetrate the immensity of its still unknown ways.

The most widely used system of measurement in the world, employed in science, technology, industry and everyday commerce is the International System of Units, abbreviation SI, also known as the modern form of the metric system. The *SI unit for time* is *Second* with *symbol **s**.*

Second is a basic unit in the International System of Units SI, one of the seven fundamental units that form the foundation for measuring various physical quantities: meter, kilogram, second, ampere, kelvin, mol, candela.

The official definition of the *second* was first given by the International Bureau of Weights and Measures / Bureau International des Poids et Mesures BIPM, at the 13th General Conference for Weight and Measures in 1967 year as:

" The second is the duration of 9192631770 periods of the radiation corresponding to the transition between the two hyperfine levels of the ground state of the caesium 133 atom. "

At its 1997 meeting, the BIPM added to the previous definition the following specification:

" This definition refers to a caesium atom at rest at a temperature of 0 K."

The BIPM restated this definition in its 26th conference in 2018 year:

" The second is defined by taking the fixed numerical value of the caesium frequency Δv_{Cs}, the unperturbed ground-state hyperfine transition frequency of the caesium 133 atom, to be 9 192 631 770 when expressed in the unit Hz, which is equal to s^{-1}. "

The *SI unit for frequency* is *Hertz* with *symbol Hz* and is defined as being equivalent to one event per second.

Hertz is a derived unit in the International System of Units SI, whose formal expression in terms of SI basic units is s^{-1}, meaning that one Hertz is one per second or the reciprocal of one second.

The Hertz is named after the German physicist Heinrich Hertz (1857-1894) who made important scientific contributions to the study of electromagnetism. The name was established by the International Electrotechnical Commission IEC in 1935 year.

The frequency of a phenomenon is determined by counting its number of periods/cycles in a second.

HEINRICH RUDOLF HERTZ

The standard units for time and frequency cannot be kept locked to have a reference always at hand. They have to be generated for each particular situation and compared with the primary standards.

Atomic clocks are built around the fundamental property of atoms to radiate at very precise frequencies.

In the 1930s the physicist Louis Essen developed the first quartz ring clock, the most accurate timepiece of its day, and a precursor of the caesium clock. Quartz clocks exploit the fact that quartz crystals vibrate at a very high frequency if the right electrical charge is applied to them. This is known as a resonant frequency, everything on earth has one.
Quartz plays the same role as a pendulum, just a lot quicker: it vibrates at a resonant frequency many thousands of times a second.
And that's where caesium atom comes in. It has a far higher resonant frequency even than quartz - 9,192,631,770 Hz, to be precise. This is one reason Essen used it to make the first of the next generation of clocks - the "atomic" clocks.
Essen's quartz creation erred just one second in three years. His first atomic clock created at National Physical Laboratory NPL in 1955 was accurate to one second in 1.4 million years.
Louis Essen chose caesium, because the frequency of its transition was at the limit of what the technology of his day could measure.

The string theory ST formulated in the 1970s years was the first to describe strings of energy. According to ST, everything in universe, all particles and forces, is comprised of tiny vibrating fundamental strings.
The second of the twelve Universal Laws is the Law of Vibration.
The Law of Vibration states that everything in universe is in a constant state of movement called vibration and the speed of vibration is called frequency or that everything in the universe is in a state of motion and vibrates at its own unique frequency.
Albert Einstein (1879–1955) was a German-born theoretical physicist who stated that everything is vibration. He received the 1921 Nobel Prize in Physics for his services to theoretical physics and especially for his discovery of the law of the photoelectric effect.
A way of storing energy by vibrating at a particular frequency is called resonance.
The things do not have a single resonant frequency.
The lowest resonance frequency is called the fundamental or the first harmonic while
all higher resonance frequencies are called overtones or harmonics.
The first overtone is the second harmonic and so on.
In an equivalent electric circuit, the resonance frequency can be derived by expressing the equal value of both capacitive and inductive reactance:
$X_L = X_C \qquad 2\pi f L = 1/(2\pi f C) \qquad f_r = 1/(2\pi \sqrt{LC})$
Different atoms and molecules have different natural vibration frequencies.
For example:
Earth resonance is called Schumann resonance and has a fundamental frequency of 7.83 Hz.
Quartz crystals are cut/shaped to vibrate at specific, high precision frequencies from few kHz to 200MHz.
Often they are made to vibrate at 32,768 Hz, easy to divide to create 1Hz signal.
32,768 is a power of 2, specifically 2^{15} and with 15 flip-flops, each dividing by 2, the final output is a 1Hz signal.
The glass brakes at the resonance frequency 556 Hz.

ALBERT EINSTEIN

FREQUENCY STANDARDS

In general, the frequency standards and the clock systems based on those are used for the control and calibration at observatories, national centres for measurement of standards, physics research laboratories, stations for watching the rockets and satellites, communication systems, navigation systems, plants and radio transmission stations.

Since their precision, the atomic frequency standards are used as reference standards in important applications in national laboratories, tracking and guide stations, interferometers, navigation receivers based on direct distance measures, geophysical expertise and communication systems.

The Caesium standard is a primary atomic frequency standard in which the radiation caused by electron transitions between the two hyperfine ground states of chemical element Caesium-133 atoms is used to control the output frequency of an electronic oscillator with quartz crystal to be 9192631770 Hz.

The Rubidium standard is a secondary atomic frequency standard in which a specified transition of electrons in chemical element Rubidium-87 atoms is used to control the output frequency of an electronic oscillator with quartz crystal to be 6834682610.904 Hz.

The first clock built on atomic frequency standard with caesium by Louis Essen in 1955 year at the National Physical Laboratory NPL in the UK was promoted worldwide by Gernot M. R. Winkler of the United States Naval Observatory. As example of use is the principal clock of the Maritime Observatory in USA, one of the most precise clocks in the world, made as a result of more than a dozen of frequency standards with Caesium. This clock controls directly the distribution of time of the naval radio systems and satellites navigation.

Atomic frequency standards based on rubidium are not primary frequency standards because their frequency is affected by internal and external factors leading to drift and instability.

In consequence the Rubidium based atomic clocks are less accurate than Caesium based atomic clocks. However, they are the most inexpensive, compact and widely produced atomic clocks, used in TV stations, in cell phone base stations, in test equipment and global navigation satellite systems like GPS.

At present there are used three types of electronic oscillators on standard frequency:

- o Oscillator with quartz crystal controlled by the electromagnetic radiation of 9,192,631,770 Hz from the Caesium-133 atoms
- o Oscillator with quartz crystal controlled by the electromagnetic radiation of 6,834,682,610.904 Hz from the Rubidium-87 atoms
- o Oscillator with quartz crystal

From the three types presented above, the first is an oscillator with a primary frequency standard and the others have secondary periodicity standards.

The difference between the oscillator with primary frequency standard and the oscillator with secondary frequency standard is that the first does not need another reference to be calibrated while the second needs calibration at the time of making and also from time to time, function of the required precision.

The frequency standard with Caesium atoms radiation is a device with atomic resonance offering access to one of the invariable oscillations of Nature.

It offers a very precise frequency standard, a primary frequency standard.

The frequency standard with Rubidium, is also a device with a resonant atomic element offering access to one of the invariable oscillations of Nature.

However, it has to be calibrated by a primary standard with the result up to 100 times better precision than the best quartz standard.

The frequency standard having as resonance element only the quartz crystal is used for less severe applications which tolerate long term drift, where electronic oscillators with quartz crystal can be used as independent generators for oscillations.

The resonance frequency of the quartz oscillators is changing with the time. After an initial aging period of few days to one month, the frequency ageing speed is almost constant. After a longer time, the accumulated drift can lead to serious error so periodical tests are needed to maintain a precise frequency standard.

The stability of frequency standards is specified in two ways:

- The long-time stability refers to slow changes of medium oscillation frequency f with the time and it is expressed as the ratio $\Delta f/f$ for a certain time period.

For quartz oscillators it is used the term "aging speed" specified in "parts/day". For Rubidium standards which are more stable are specified in "parts/month". The Caesium standards are primary units without systematic drift, so the frequency of these primary standards is guaranteed for a specified precision.

- The short time stability is referring to frequency changes for a sufficient short time so the long-time stability to be neglected.

In the tests for the frequency standards are taken in consideration two variation ways, accidental or systematic and two measurement methods, in time domain and in frequency domain. Each of the measuring methods answers for both variations, accidental and systematic.

For all frequency standards, the oscillation generator is an electronic oscillator with quartz crystal (MHz).

For atomic frequency standards (Caesium and Rubidium) presented in Fig 1 the oscillation frequency of the electronic oscillator circuit with quartz crystal is multiplied to an atomic resonance frequency (6834 MHz for Rubidium and 9192 MHz for Caesium).

Then this signal is modified/modulate in frequency to pass through the atomic resonator, designed to resonate at a particular frequency.

The output signal of atomic resonator block is amplified and applied to the quartz oscillator to control its frequency.

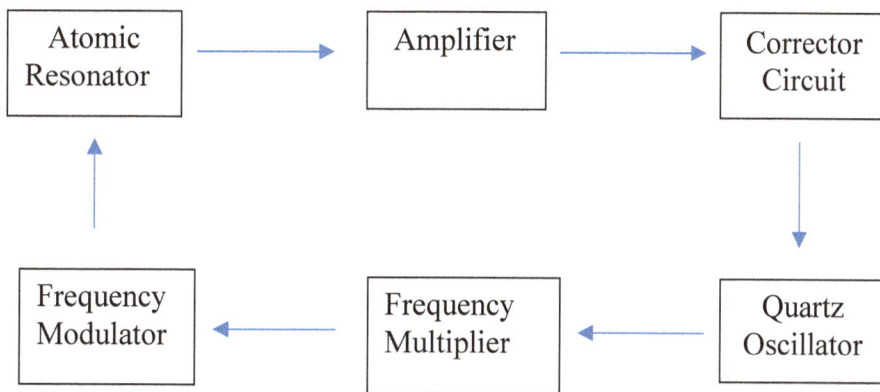

FIG 1   ATOMIC PERIODICITY STANDARD

1. The frequency standard using a resonator with Caesium beam atoms exposed to microwave radiation as stabilising element, has a standard deviation of 1×10^{-12} without calibration. It is a primary frequency standard of atomic type. Its advantages: absolute precision, intrinsic reproducibility and no perceptible drift for long periods of time.

The best standards proved their performance in over 60 million hours of functioning.

The main variety of atomic clocks uses Caesium atoms cooled to temperatures that approach absolute zero.

2. The frequency standard using a resonator with Rubidium steam exposed to microwave radiation as stabilising element is a secondary frequency standard of atomic type. Its stability is better than 1×10^{-11} per month which is 50 to 100 times better than the quartz frequency standard.

3. The frequency standard using quartz crystals is the third frequency standard type which is a performance -in time and frequency- precision system since its very good characteristics for thermal stability for long or short periods of time, capacity to work in different environments.

Typically, the oscillator drifts by 1×10^{-9} in 30 minutes after an off period of 24 hours.

New technologies created quartz crystals with lower ageing speed.

This standard is suitable for Doppler measurements, spectroscopy of microwaves, communication and navigation systems, synthesizers, time code generators, spectrum counters and analysers.

PRECISE MEASUREMENT OF TIME

From the above presentation it is clear that a precise measurement of time
it is possible using precise frequency oscillators, manly based on atomic
oscillations.

The best measurements of time, the best clocks in world represent a happy
joint of the properties of the atomic and molecular structures and the properties
of the electrical circuits, being the result of collaboration between physicians
and electronicists.

The precision is 1 second in few thousand years, if we take in consideration
large time intervals and is a thousand times bigger for short time intervals.

The block diagram of a digital precision clock is presented in FIG 2.

It comprises:

- frequency standard signal followed by digital impulse generator, an analogue
  to digital converter ADC
- frequency divider giving periodic electric impulses with frequency $f = 1$ Hz
  and period $T = 1$s (digital counter realised for example with IC 7490)
- binary counter of impulses which come at each second and so recording
  the time (digital counter realised for example with IC 7490)

Because counting is in binary code and the time display is in decimal code,
follows:

- system to translate numbers from binary code representation to decimal code
  representation
- digital display in decimal code

The clock precision is determined by the oscillator on standard frequency.

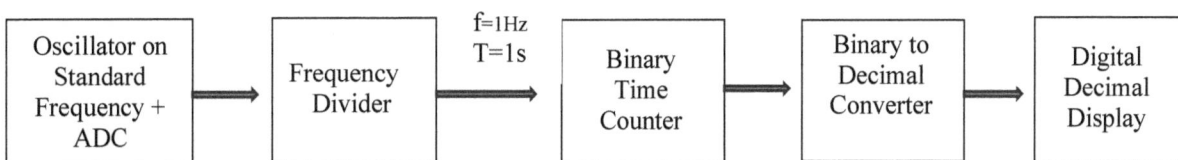

Oscillator on Standard Frequency + ADC	Frequency Divider	$f=1Hz$ $T=1s$ Binary Time Counter	Binary to Decimal Converter	Digital Decimal Display

Fig 2   BLOCK DIAGRAM OF PRECISION CLOCK - TIME METER

THE TIME METER – THE CLOCK

Digital clocks are very common nowadays and their block and schematic diagrams are well known. Further it is presented an example, the design of a digital clock by the way of block diagram of Fig 2. The oscillator is with quartz crystal resonance on frequency 1MHz.

QUARTZ CRYSTAL OSCILLATOR & IMPULSE GENERATOR

Following the block diagram for the precision clock of Fig 1 we observe that the first block is an oscillator on standard frequency. Its output analogue oscillations will enter a digital impulse generator, which is an analogue to digital converter ADC, since the here described clock is a digital device.

The block can be realised with an electronic oscillator with quartz crystal with a resonance frequency of 10^6 Hz, followed by an electronic circuit generator of positive electric impulses, see Fig 4.

The symbol and equivalent electrical circuit of the quartz crystal is presented in Fig 3.

The stability with quartz is known as a basic method to obtain a high frequency stability of the generated oscillations.

An essential progress in using on large scale the quartz oscillators was obtained first by producing quartz crystals with cut AC (biconvex and plan convex), which have the best monochromaticity, a very high-quality factor $Qc = (1 \div 3) \times 10^6$ and a low temperature coefficient.

The frequency variation due the ageing of quartz at small oscillation amplitudes (20mV) is $(1 \div 5)10^{-7}$/year.

Functioning of transistor with small supply voltage implies a low voltage applied on quartz. It was succeeded to reduce at minimum the destabilizing action of transistor parameters by using a transistor with a limit frequency f_s at least ten times higher than the work frequency, by connecting resistors in the emitter and base circuits and mainly by loosing seriously the connection with the quartz, possible due to the big value of the transistor slope S characteristic ($S = 200 \div 500$mA/V).

The here presented oscillator realized with the transistor T_1 respects the above conditions.

T_1 is a BF 183 NPN bipolar transistor, a 25V 15mA RF-IF high-frequency 800 MHz transistor in TO-72 case, from Digi-Key.

The direct voltage applied on base varies between 0.99V and 2.96V resulting a direct current between 0.45mA and 1.35mA. The collector voltage varies between 4.45V and 1.49V.

The oscillations amplitude is function of the base voltage in consequence function of the collector current.

Transistor T2 is in emitter follower connection, to separate the oscillator block from the digital impulse generator block realised with the transistor T3.

T_2 and T_3 are AC 181 NPN bipolar Germanium junction transistor BJT in T01 package

P_c 0.3W V_{CB} max 32V V_{CE} max 16V I_c max 1A   Op Junc Temp Max 100 °C

V_{EB} max 20V   Transition Frequency F_T 2MHz   Collector Capacitance C_C 60 pF from UXPython inc.

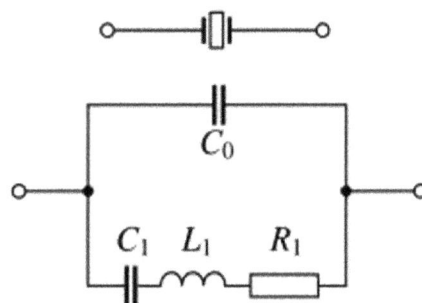

FIG 3   SYMBOL OF QUARTZ CRYSTAL
ELECTRICAL EQUIVALENT OF QUARTZ CRYSTAL

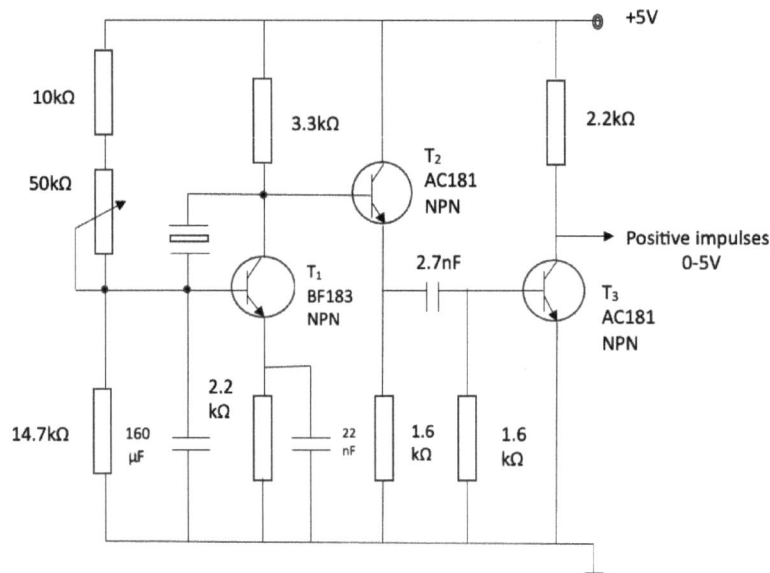

FIG 4   1 MHz QUARTZ OSCILLATOR & DIGITAL IMPULSE GENERATOR

STANDARD FREQUENCY DIVIDER

The 1MHz impulses of 0-5V amplitude from the T3 collector enter the frequency divider built with 6 integrated circuits IC 7490, each performing a reduction by 10 of frequency.

Its aim is to obtain impulses with the frequency f = 1Hz and period T = 1s.

They are counted in the next block which is actually the counter of time, the clock.

Integrated circuit IC 7490 is a Decade Counter, counting in BCD code.

In computing and electronic systems, binary-coded decimal BCD is a class of binary encodings of decimal numbers where each decimal digit 0-9 is represented by a fixed number of bits, usually four or eight, Fig 5.

| COUNTING | | OUTPUT | | |
	D	C	B	A
0	0	0	0	0
1	0	0	0	1
2	0	0	1	0
3	0	0	1	1
4	0	1	0	0
5	0	1	0	1
6	0	1	1	0
7	0	1	1	1
8	1	0	0	0
9	1	0	0	1

FIG 5   COUNTING IMPULSES IN BCD CODE REPRESENTED BY 4 BITS

Integrated circuit IC 7490 is a Decade Counter, manufactured originally by Fairchild Semiconductor, see Fig 6. It is an integrated circuit used in digital electronics for impulses counting and sequencing purposes, with applications in electronic systems, useful in digital counting, frequency division and timing applications.

Developed as a monolithic device, IC 7490 contains a combination of flip-flops that have two stable states that can store information/digits and digital logic gates that allow it to count in variety ways.

IC 7490 is built with 4 flip-flops A B C D, with the outputs A B C D, so interconnected to realise two frequency dividers, a divider by 2 and a divider by 5.

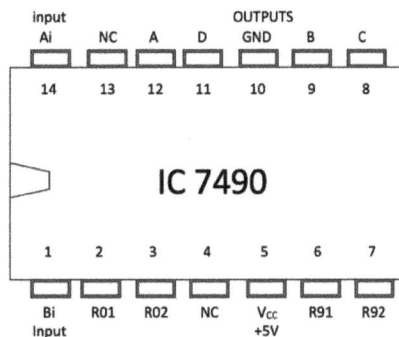

FIG 6   DECADE COUNTER IC 7490

69

Its inputs R01 R02 and R91 R92 permit or inhibit the counting of impulses and also bring all outputs in 0 state or in the state to represent in binary code the digit 9, see Fig 7.

INPUTS INITIALIZATION				OUTPUTS			
R01	**R02**	**R91**	**R92**	**D**	**C**	**B**	**A**
1	1	0	X	0	0	0	0
1	1	X	0	0	0	0	0
X	X	1	1	1	0	0	1
X	0	X	0	COUNT			
0	X	0	X	COUNT			
0	X	X	0	COUNT			
X	0	0	X	COUNT			

FIG 7   INPUTS INITIALIZATION FOR IC 7490

Because the output A is not connected internally, the integrated circuit IC 7490 can function:

- To divide by 2 and to divide by 5 separately for what are not needed external connections.

The flip-flop A is used as binary element for the division by 2 with input A_i and output A.

Input Bi is used to obtain the division by 5 at outputs B C and D.

So, the two counters function independently; however, the 4 flip-flops are brought to zero 0 simultaneously.

- To divide by 10

Used as decade counter, the input Bi must be connected to output A.

The counting impulses are applied on Ai, the input of flip-flop A, and the counting sequence is in conformity to the table for counting in code BCD, see FIG 5.

TIME COUNTER FOR SECONDS, MINUTES AND HOURS

This counter counts in BCD code the impulses which come at each second 1s.
It is the time counter.

It is built with 6 integrated circuits IC 7490:

Seconds counter, built with first IC 7490 (counts ten seconds) and second
IC 7490 (counts sixty seconds with feedback), giving impulses at each
60 s = 1 minute:

> First IC 7490 divides by 10
>
> Second IC 7490 divides by 6 with feedback, giving at output impulses
> with T=1min - see FIG 8 with detailed connections only for feedback

Minutes counter, built with third IC 7490 (counts ten minutes) and fourth
IC 7490 (counts sixty minutes with feedback), giving impulses at each
60 minutes = 1 hour:

> Third IC 7490 divides by 10
>
> Fourth IC 7490 divides by 6 with feedback, giving at output impulses
> with T=1hour - see FIG 8 with detailed connections only for feedback

Hours counter, built with fifth and sixth IC 7490 (counts 24 hours with
feedback), giving impulses at each 24 hours:

> Fifth and Sixth IC 7490 divide by 24 with feedback, giving at output
> impulses with T=24h - see FIG 9 with detailed connections only for feedback

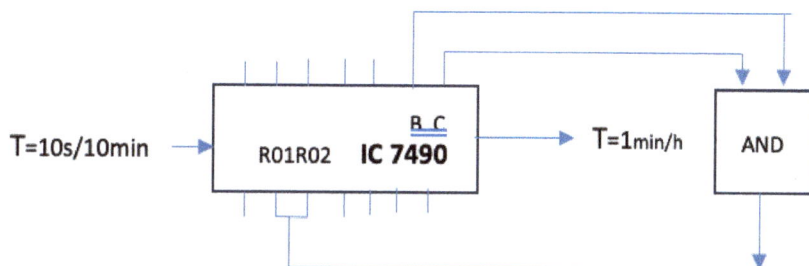

FIG 8   IC 7490 WITH FEEDBACK CIRCUIT TO DIVIDE BY 6

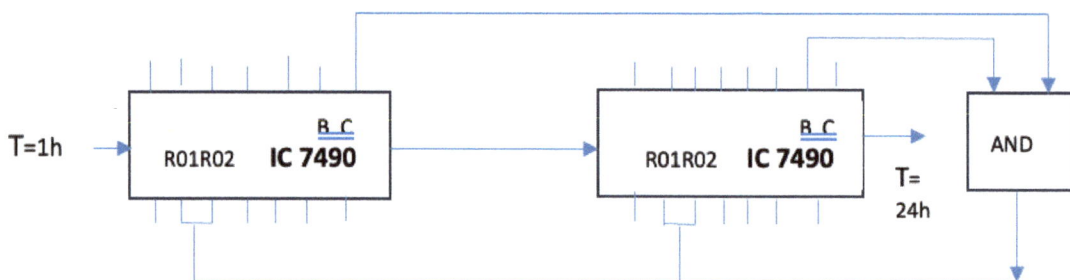

FIG 9   TWO IC7490 WITH FEEDBACK CIRCUIT TO DIVIDE BY 24

BINARY-DECIMAL DECODER

The counting of the seconds, minutes and hours in the TIME COUNTER is in binary code.

The display of minutes and hours is on NIXIE tubes in decimal code.

The translation of numbers from binary code representation to decimal code representation it is done with integrated circuit IC 7442, the binary-decimal decoder seen in FIG 10.

IC 7442 has four inputs corresponding to outputs A B C D of Decade Counter IC 7490 and ten outputs corresponding to digits 0 1 2 3 4 5 6 7 8 9, where for each digit appears a negative impulse.

These binary-to-decimal decoders consist of eight inverters and ten, four-input NAND gates. The inverters are connected in pairs to make BCD input data available for decoding by the NAND gates.

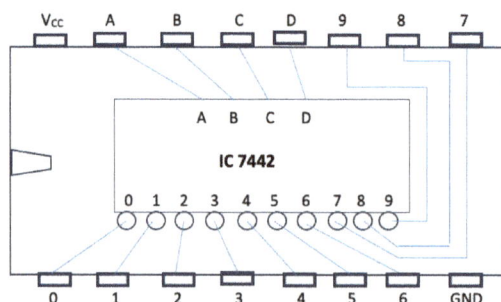

FIG 10   BINARY-DECIMAL DECODE IC 7442

72

NIXIE TUBES AND THEIR DRIVERS

Each tube NIXIE contain an anode and 10 cathodes in the shape of the digits 0,1,2,3,4,5,6,7,8,9.

The NIXIE tube is a tube with gas. To turn it on, is necessary a voltage 180V in series with a resistor 100 kΩ to limit the current (the voltage on tube is 130V in functioning at a current 0.5mA per digit).

The tube anode is connected to the positive terminal of the 180V electric power supply and each cathode is connected to the collector of a transistor T with the emitter at ground.

T is a BC107 transistor, silicon planar epitaxial NPN bipolar transistor, low power with

$U_{max} = 45$ V $I_{max} = 200$ mA, in TO-18 metal case from ST Microelectronics, widely used in various driver applications. When on the transistor base appears a positive impulse, the transistor enters in conduction and the NIXIE tube receives the necessary current to display a digit (the digit of the tube cathode connected in the collector of the respective transistor) see FIG 11.

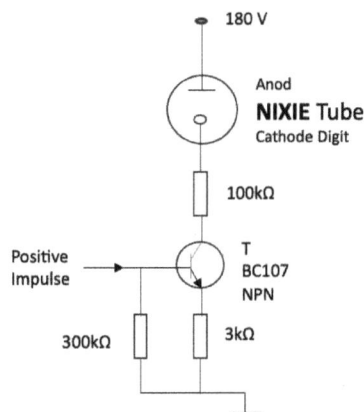

FIG 11   DRIVER FOR NIXIE TUBE

Since the binary-decimal decoder gives negative polarity command impulses and the drivers should be driven by positive polarity impulses, between them are introduced inverters.

Are used integrated circuits IC 7404, each having 6 inverters, see FIG 12.

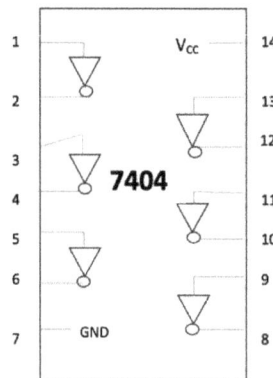

FIG 12   HEX INVERTER IC 7404

The 7404 IC, also known as the hex inverter, is comprised of six independent inverters, each capable of transforming logic high signals to logic low and vice versa. Its versatility makes it an indispensable building block for numerous applications, ranging from basic logic gates to more complex digital circuits.

SCHEMATIC DIAGRAM OF THE TIME METER

The logical connections of the designed electronic time meter or clock having an internal oscillator on standard frequency realised with quartz crystal is presented in FIG 13.

It is observable:

- the quartz oscillator of 1 MHz, + digital impulse generator ADC

- the frequency divider by 10^6 realised with six Decade Counters IC 7490

- the counters for seconds, minutes, hours each built by two IC 7490

- the binary-decimal converter/decoder built with four IC 7442

- the drivers or circuits to attack four NIXIE tubes

- the four NIXIE tubes for displaying the minutes and hours in decimal code

DIGITAL TIME METER - CLOCK

S = Switch with 2 positions T & U
T = Time Meter - Clock
U = Update Time with 2 pos. for minutes & hours

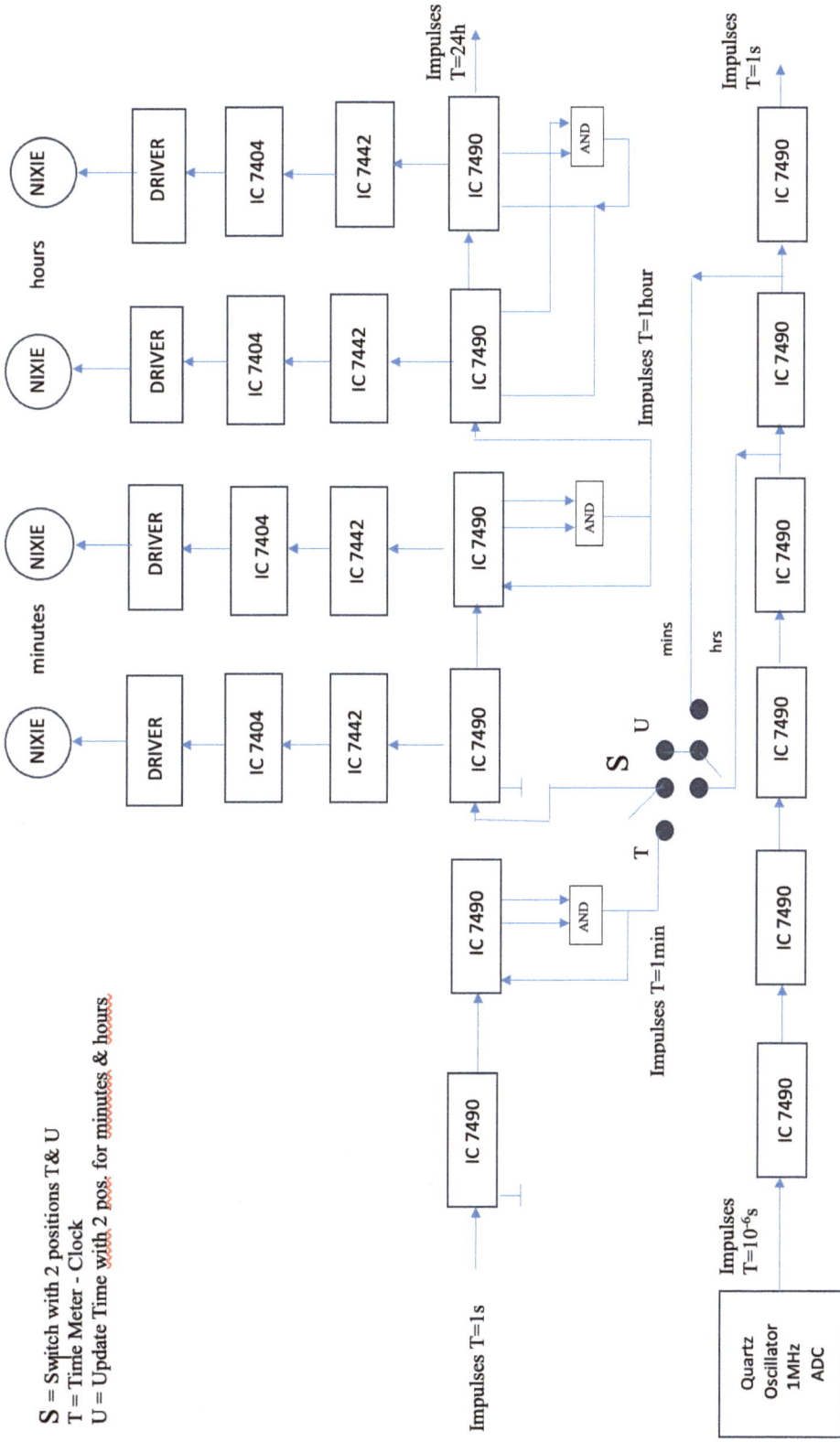

FIG 13

PRECISE MEASUREMENT OF FREQUENCY

The duality *time - frequency* is materialized in the production of electronic devices in the duality *time meter (clock) - frequency meter.*

Various types of mechanical frequency meters were used in the past, but since the 1970s years these have been almost universally replaced by digital frequency counters.

A digital frequency meter is a digital counter to count the number of rising or falling signal edges occurring in the measured signal within a specific period of time, known as gate time.

The best electronic frequency meter on international market in 1979 year, produced by Hewlett-Packard has a resolution of $2x10^{-9}$ independent of the input frequency, so 1MHz could be measured with a precision of 0.002Hz/s, while the conventional frequency meter offered a precision of 1 Hz, 500 times lower.

Nowadays, depending on the specific instrument, the frequency meters can measure a vast range of frequencies from few Hertz up to several gigahertz GHz with 0.01 Hz resolution, making them acceptable for several applications, from power grids to high frequency telecommunications.

In older days, a luxury in big metrology laboratories and few crystals factories, the digital frequency meter is now common in laboratories, production lines, being a precision tool in the automatic systems.

More, they became more multilateral with larger applications.

For example, it is known that the quartz crystals resonate on very precise frequencies. Was discovered a way to produce quartz crystals whose resonance frequency varies very linear with the temperature or pressure for example with 1000 Hz at every Celsius degree.

In that way, by measuring the resonance frequency, can be done a very precise temperature or pressure measurement.

So, there is a standard precision of 0.0001 grad Celsius in the range -80 °C ÷ 250 °C using this kind of temperature translator – the quartz crystal with the digital frequency meter.

The frequency measurement is a fundamental measurement which is realised adding the number of input cycles or events in a known period of time, called *gate time*.

The total resulted number is or is proportional with the unknown frequency. Usually, the gate time is one second 1s.

The input voltage signal whose frequency should be measured is applied - through an input circuit ADC which changes it in digital electric impulses if it is a wave - to a counter circuit which counts the coming digital impulses in time of a second.

The block diagram of a precision frequency meter is presented in FIG 3.

It comprises six main blocks.

They are:

1. oscillator on standard frequency and ADC (analogue to digital converter)
2. standard frequency divider (change standard frequency to 1Hz frequency)
3. input circuit + ADC
4. frequency counter
5. binary-decimal decoder
6. digital frequency display

○ The first block is an oscillator on standard frequency, followed by an analogue to digital converter ADC which transforms the oscillating waves in digital electric impulses.

The best standard frequency is given by the quartz oscillator controlled by Caesium atoms radiation of 9,192,631,770 Hz.

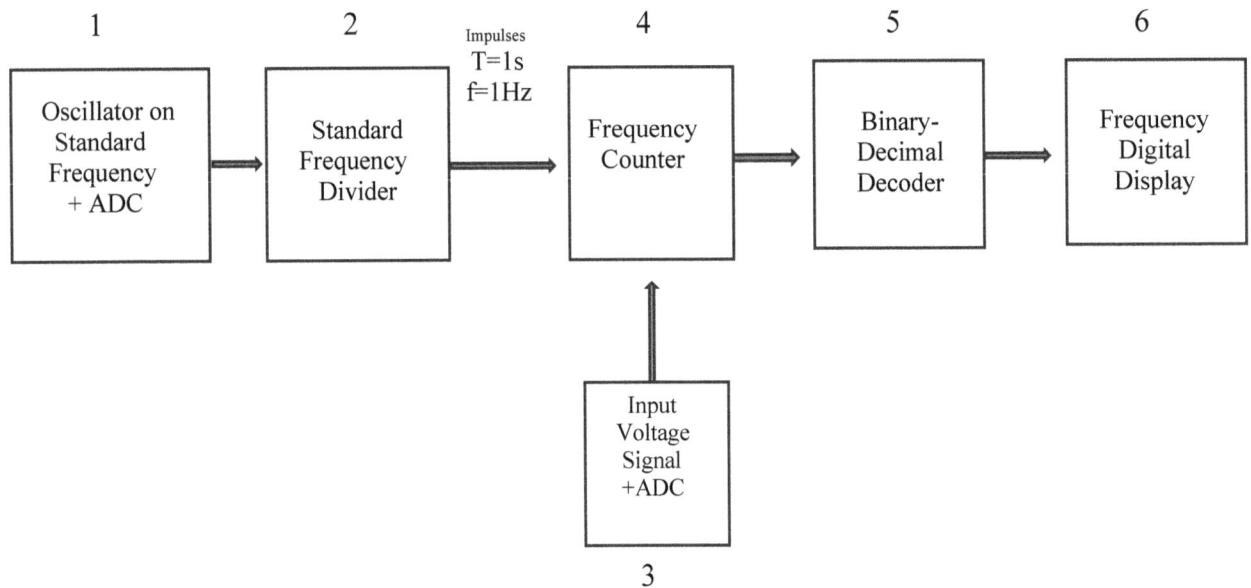

```
   1              2       Impulses     4              5              6
                           T=1s
                           f=1Hz
┌──────────┐  ┌──────────┐      ┌──────────┐  ┌──────────┐  ┌──────────┐
│Oscillator on│ │ Standard │      │Frequency │  │ Binary-  │  │Frequency │
│ Standard │  │Frequency │ ───> │ Counter  │  │ Decimal  │  │ Digital  │
│Frequency │  │ Divider  │      │          │  │ Decoder  │  │ Display  │
│  + ADC   │  │          │      │          │  │          │  │          │
└──────────┘  └──────────┘      └──────────┘  └──────────┘  └──────────┘
                                      ▲
                                ┌──────────┐
                                │  Input   │
                                │ Voltage  │
                                │ Signal   │
                                │  +ADC    │
                                └──────────┘
                                     3
```

FIG 3   BLOCK DIAGRAM OF PRECISION FREQUENCY METER

- The second block is a frequency divider (digital counter realised for example with integrated circuits IC 7490) which reduces the frequency of digital impulses to 1Hz and period T=1s.

- The third block is the input circuit for the input signal whose frequency should be measured (contains an analogue to digital converter ADC in case the input signal is wave).

- The fourth block is the frequency counter (digital counter realised for example with integrated circuits IC 7490) which counts in the time of a second the impulses from the input circuit whose frequency should be measured.

  The number in counter at the end of the second represents the number of impulses in a second, in consequence the frequency of input signal. That number is maintained in counter for the next second for display. At the end of the second second, the counter is brought to zero, deleted and the counting begins again.

o The fifth block is a binary-decimal decoder which does the translation of numbers from binary code representation to decimal code representation (realised for example with integrated circuit IC 7442, the binary-code decimal decoder).

o The sixth block is the digital display of frequency, realised for example with NIXIE tubes.

references

Electronic Time Measurements – book edited by Britton Chance, Robert I. Hubsizer, Edward F. MacNichol, Frederick C. Williams - 1949

Time Measurement / An Introductory History – book by Kenneth F. Welch 1972

Electronic Instruments and Systems   Hewlett Packard   1979

The Measurement of Time / Time, Frequency and Atomic Clock – book by Claude Andoin & Bernard Guinot 2001

Galileo's Pendulum / from the Rhythm of Time to the Making of Matter - book by Roger G. Newton 2004

The top 6 reasons to use silicon MEMS timing solutions – EDN article by Piyush Sevalia & Aaron Partridge June 2014

A Brief History of Timekeeping / How Humans Began Telling Time – video 2019

Focus on Timing - article by Harvey Toyama, Patrick Mannion, Daniel Bogdanoff, John Blyler editor Bill Wong / Electronic Design Library 2020

Measurement of Time – video 2020

Who decides how long a second is? John Kitching – video 2021

Timing Decisions 102: Optimize Your Clock Tree – article by James Wilson / Electronic Design Oct 2023

How humanity has measured time: From prehistory to the digital age – video 2025

Digi-Key Corporation American electronics component distributor– Microchip integrated circuits

Fairchild Semiconductor – Datasheet

Ovaga Technologies

STMicroelectronics

Wikimedia   Wikipedia

THERMOREGULATOR WITH THERMOCOUPLE

The temperature is a state function, that characterises the matter from the heat view point.

Temperature is the degree or intensity of heat present in a substance or object, especially as expressed according to a comparative scale and shown by a thermometer as sensor/meter or perceived by touch.

Since the heating degree is a determinant factor for many properties of materials, the knowledge about it interested much the specialists, leading to the creation of many devices for sensing/measuring the temperature. Electrical and electronic devices have permitted the measuring, recording, transmission at distance and the automatic control of temperature.

The development of semiconductors industry facilitated the creation of new devices characterised by simplicity, autonomy, high precision, safe to use etc. The devices for the measurement of temperature using thermocouple as sensor are very spread in all fields of human activity because they are simple, safe to use and have high precision and low price, also give good results in the controlling of temperature in automated installations.

Thermocouple is a sensor/device for sensing/measuring the temperature. It comprises two dissimilar metallic wires joined together to form a junction. When the junction is heated or cooled, a small voltage is generated in the electrical circuit of the thermocouple. The value of the small voltage is function of the temperature. This is the thermoelectric effect discovered by the Baltic German physicist Thomas Johan Seebeck in the 1822 year and interpreted further by French physicist Jean Charles Athanase Peltier and Lord Kelvin. Measuring the voltage, it is measured the temperature.

The materials used to make thermocouples are:
- for temperatures under 1000°C – Metals and Alloys as:
  Iron Fe, Copper Cu, Constantan (55% Copper Cu + 45% Nickel Ni),
  Chromel (90% Nickel Ni + 10% Chromium Cr),
  Alumel (95% Nichel Ni + 2% Aluminium Al + 2% Manganese Mn + 1% Silicon Si),
  Copel (56% Copper Cu+ 44% Nickel Ni), Chromium Cr etc.
- for temperatures between 1100°C and 1600°C – Noble metals as:
  Gold Au, Silver Ag, Platinum Pt, Thorium Th, Iridium Ir etc.
- for temperatures over 1600°C – Refracting materials as:
  Wolfram (Tungsten) W, Molybdenum Mo, Silicon Carbide SiC etc.

The further described thermoregulator uses the thermocouple as temperature sensor and the thermocouple is of type Chromel - Alumel.

The thermoregulator is used to heat an oven and control the oven temperature.

In principle, the voltage supplied by the thermocouple is compared, using an operational amplifier, with the voltage supplied by a potentiometer wiper.

A variable resistor, known as potentiometer/rheostat allows you to adjust the amount of resistance in an electric circuit. A potentiometer or "pot" is a three-terminal resistor with a sliding/rotating contact that can form an adjustable voltage divider, highly useful in a wide range of applications. The wiper of a potentiometer is a sliding contact that moves along a resistive element to change the resistance in a circuit.

The voltage difference is amplified by an operational amplifier and used to command the electric current which heats an oven in such a way that the oven temperature should increase to the value which would create in the thermocouple a voltage equal with that supplied by the pot wiper.

So, it is possible to arrange the oven temperature by the position of the wiper of a potentiometer. The potentiometer can be calibrated in temperatures. The electronic schematic diagram of the thermoregulator is presented in fig 1.

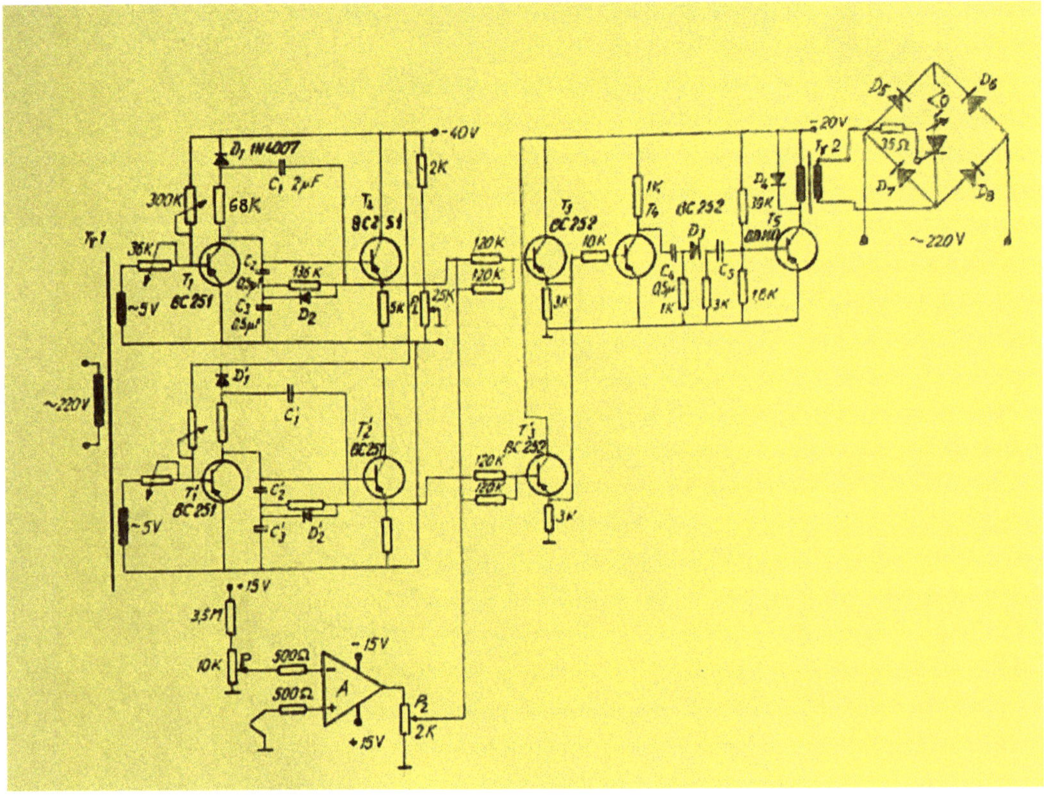

Fig 1

Mains electricity or powerline used in design is a general-purpose alternating-current AC electric power supply with standard voltage 220 V, frequency 50 Hz.

The voltage generated by the thermocouple is compared with the voltage supplied by the wiper of potentiometer P, voltage corresponding to the desired temperature.

The voltage difference amplified by amplifier A, together with the positive sawtooth wave generated by the transistors T_1T_2 respectively $T'_1T'_2$, will turn ON / OFF the transistor T_3 respectively T'_3.

If the output of the operational amplifier is negative voltage, that will switch ON the transistor T_3 respectively T'_3 (in the time of powerline semi-wave).

If at the output of the operational amplifier is a positive voltage the transistor T_3 respectively T'_3 will be OFF.

When the transistor T_3, respectively T'_3 is ON, the transistor T_4 is also ON and the positive voltage variation on its collector will cut OFF the transistor T_5; the current variation in T_5 creates a positive voltage variation in the secondary of transformer T_r2, which will turn ON the thyristor T supplied by powerline through a bridge of diodes D5 D6 D7 D8.

The thyristor T is in serial connection with the heating wire of the oven.

When T is ON an electric current will pass through heating wire of oven, heating the oven at the desired temperature.

Input Transformer

The input transformer Tr1, see fig 7, assures to obtain two voltages 5 V_{ef} with the powerline frequency 50 Hz and in antiphase one to the other.

A transformer is a device utilized for electrical power transmission via electromagnetic induction.
Electric power is transferred without any frequency or phase modification.
A basic transformer is made of two coils of wire; a primary coil from the ac input and a secondary coil leading to the ac output. The coils are not electrically connected. Instead, they are wound around an iron core. This is easily magnetised and can carry magnetic fields.

A power transformer consists of various unique components, each contributing differently to the overall performance of the transformer: core, windings, insulating materials, transformer oil, breather, cooling tubes and others.

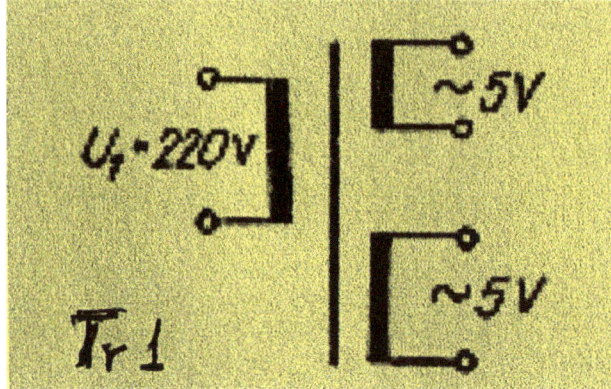

Fig 7

Considering the base current of the transistor T_1 in the range of tens of μA, the section of the transformer core is determined by the existing tola.

It is chosen the transformer core strip E8 giving the square section of core.

Results the number of winds per volt for Tr_1 primary n' and Tr_1 secondary n'' considering the induction: $B_{max} = 11 \cdot 10^3$ Gs = 11 kGs.

$n' = 10^8 / (4.44 \cdot B_{max} \cdot S_m \cdot f) = 16.1$/volt

$n'' = 1.1 \cdot 16.1 = 17.7$/volt

$U_1 = 220$ V $\qquad n_1 = U_1 \cdot n' = 220 \cdot 16 = 3520$ winds

$U_2 = U_3 = 5$ V $\qquad n_2 = n_3 = U_2 \cdot n'' = 5 \cdot 18 = 90$ winds

Ø cooper wire = 0.05 mm

Should be assured the T_1 base polarisation of approximately 0.5 V by the resistor 300 kΩ.for T_1 opening in saturation, having the resistor serial with Tr_1 transformer secondary of approximately 20 kΩ.

It is preferable to use variable resistors for easy adjustments.

86

Sawtooth Wave Generator

The sawtooth wave or saw wave is a kind of non-sinusoidal waveform. It is so named because resemblances the teeth of a plain-toothed saw. A single sawtooth is called a ramp waveform.
The sawtooth wave is actually repeated linear variable voltage.

In the time of a period of the mains electricity (220V 50Hz) are generated two linear variable voltages (sawtooth wave or saw wave) in the circuit formed by the transistors T_1T_2 respectively $T'_1T'_2$.

The sawtooth wave at the emitter repeater T_2 (T'_2) varies between +15 V ÷0 V.

The sawtooth wave is a repeated linear variable voltage.

In principle, to obtain a voltage u with a linear variation, it is used the charging of a capacitor under constant current I, generated by a current generator I, see fig 2.

In electrical engineering a capacitor is a device that stores electrical energy by accumulating electric charges on two closely spaced metallic surfaces/plates that are insulated from each other by a dielectric medium: glass, ceramic, plastic film, paper, mica, air, oxide layers with thickness d.
When an electric potential difference, a voltage V, is applied across the terminals of a capacitor, an electric field E develops across the dielectric, causing a positive electric charge +Q to collect on one plate and negative electric charge –Q to collect on the other plate.
An ideal capacitor is characterised by a constant capacitance C with the measuring unit the farad F.
The capacitance is defined as the ratio of the positive/negative electric charge Q on each plate to the voltage V between them: C = Q/V

$V = \int_0^d E(z)\, dz = Ed = sd/e$
$E = s/e = $ electric field $s = Q/A = $ charge density $e = $ electric permittivity of dielectric
$V = Qd/eA$ $C = eA/d$

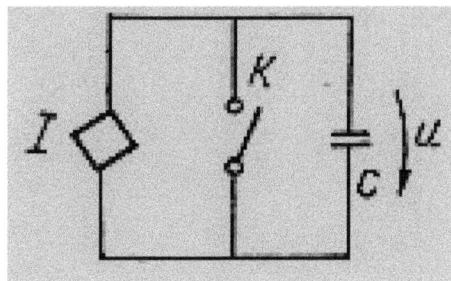

Fig 2

The voltage at the capacitors terminals varies linear in time:

$u = (\int i(t)\, dt + U_C(0))/C = It/C$

$U_C(0) = $ initial voltage at the capacitor terminals, here considered zero.

In case that for charging the capacitor, it is used a voltage generator U with the internal resistance R it is necessary a linearisation.

The procedure for linearisation is:

It is introduced an additional voltage supply e serial with the charging circuit of the capacitor C, the additional voltage varying so that the charging current of the capacitor C is maintained constant, see fig 3.

To maintain constant the charging current I, should be accomplished the condition:

$I = (U + e - u) / R = $ constant

Will result $e = u$, meaning that the voltage u variation to be followed by the voltage e.

For $e = u$ $I = U/R = $ constant   and $u = Ut / RC$

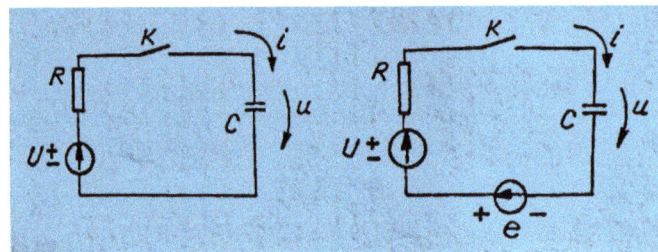

Fig 3

Considering the circuit in fig 4 it is observed that it accomplishes the condition:

$u = Ut / RC$ for $A = 1$

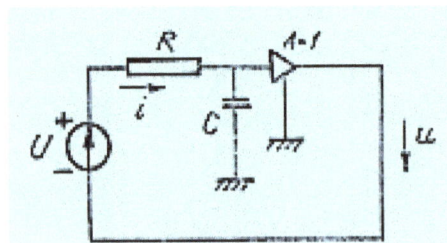

Fig 4

A practical circuit corresponding to the example of fig 4 is given in fig 5.

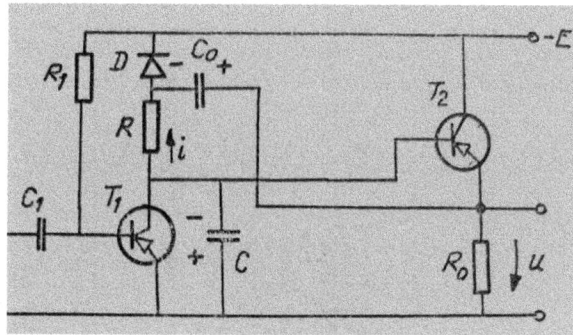

Fig 5

At rest the transistor T_1 is in conduction at saturation and maintains the capacitor C uncharged, so this transistor plays the role of commutator K.

In these conditions the capacitor C_0 of big capacity $C_0 \gg C$ is charged at a voltage close to that of the voltage supply E, charging done through diode D.

The capacitor C_0 plays the role of voltage supply U.

Applying a positive impulse on the T_1 base, this transistor will be OFF and the capacitor C is charging through the resistance R from the supply U represented by C_0.

In the same time, serial with U appears the voltage u from the T_2 emitter.

That voltage reproduces the voltage at the terminals of capacitor C which is applied on the T_2 base.

So, the linearisation conditions are accomplished and the voltage u is increasing linearly.

The active go is determined by the time period of the command impulse (the positive demi-wave taken in the Tr_1 transformer secondary supplied in its primary by the powerline).

At the end of this impulse the transistor T_1 is ON and the capacitor C discharges fast through the T_1 saturation resistance which has small value.

In literature, it is given a similar circuit of a superior voltage linearity of saw wave shown in fig 6.

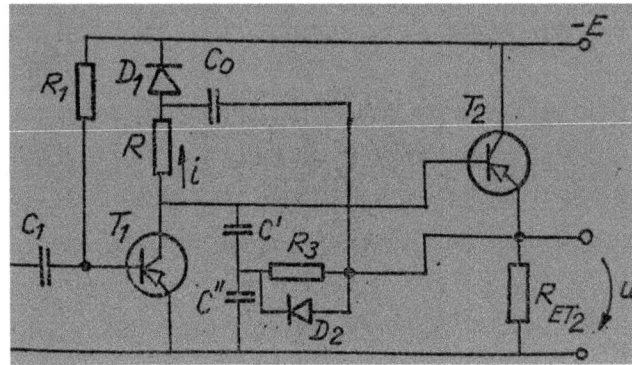

Fig 6

Analogue with the circuit of fig 5, the circuit from fig 6 contains more:

the resistor R_3 of value $R_3 = R / 4(C/C_0 + 1 - A_{T3})$, the diode D_2 and

the capacitor C is divided in two capacitors of equal capacity $C' = C'' = 2C$.

In the thermoregulator electronic schema of fig 1 the supply voltage E is 40 V.

The transistors T_1 T'_1 T_2 T'_2 are of type BC 251.

The diodes D_1 D_2 are of type EFD 108.

$C_0 = 2\mu F$

$C' = C'' = 0.5\ \mu F$

$R = 68\ k\Omega$

$R_3 = 136\ k\Omega$

$R_{ET2} = 5.6\ k\Omega$

$R_1 = 360\ k\Omega$

Time constant of capacitor
Time constant τ is a concept in electrical engineering, measuring the response time of a system to a step input.
For an RC circuit $\tau = RC$.
For an RC circuit, after one time constant, the circuit reaches approximately 63.2% of its final value of the step input, not 100%. Only after 5τ the capacitor is charged to 100% of final value of the step input.

90

The time constant τ of system formed by capacitor C (C' in series with C") and resistor R for R=68 kΩ is:

$\tau = R C = R C' C'' / (C'+C'') = 68 \times 10^3 \times 0.25 \times 10^{-6}$ s $= 17 \times 10^{-3}$ s $= 17$ ms

The charging of C is a saw tooth wave.

After time τ the capacitor C is charged 63.2% of E, that is

63.2 x 40 V = 25.28 V

After 10 ms the capacitor C is charged at 10 x 25.28 / 17 = 14.9 V ~ 15 V

The voltage u on R_{ET2}, which repeats the charging of C is a saw tooth wave.

In consequence the voltage u on R_{ET2} is a saw tooth wave with variation

$0 \div -15$ V.

The sawtooth voltage obtained on resistor R_{ET2}, respective $R_{ET'2}$, evolves like in fig 8.

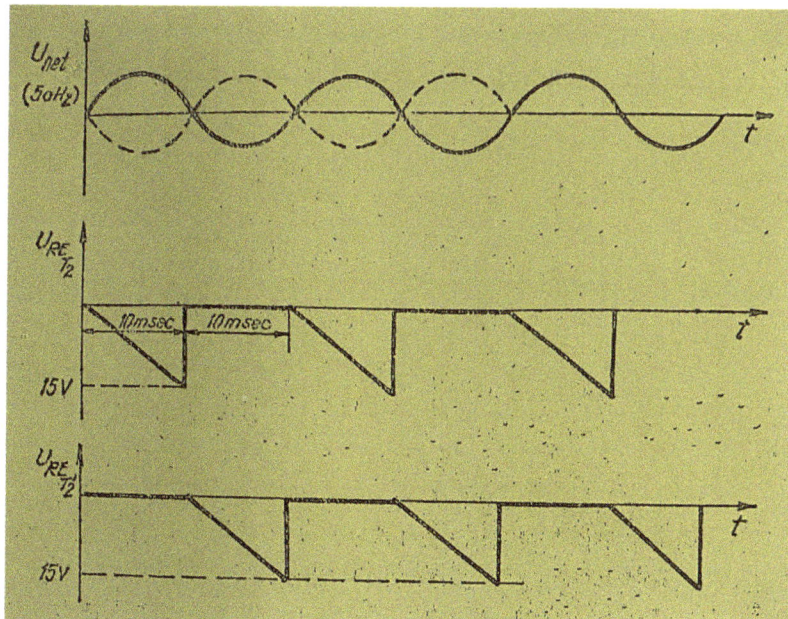

Fig 8

The block time for the transistor T_1 is:

$T_{powerline} / 2 = 20$ ms / 2 = 10 ms

It is considered the voltage u on R_{ET2} ($R_{ET'2}$)

$u = 0 \div -15$ V

With the help of variable resistor P1 the variation $0 \div -15$ V is converted to $+15$ V $\div 0$ V, see fig 9.

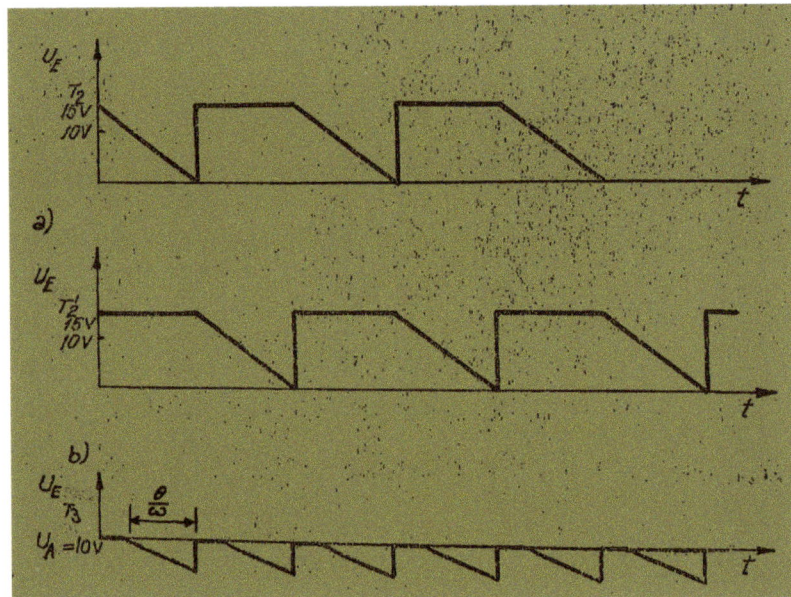

Fig 9

Opening Impulse of Thyristor

The transistor T_3, respective T'_3, type BC 252, together with its emitter resistance is an emitter repeater.

$R_E = 3$ kΩ

$R_{inputT3} = h_{11e} + (\beta + 1)R_E = 607.5$ kΩ

The transistor T_4, also type BC 252, is usually OFF, its collector voltage is -20 V.

The capacitor C_4 is charged so long the transistor T_4 is OFF.

$C_4 = 0.5$ μF

C_4 charges through two resistors 1kΩ, with $\tau = 2 \cdot 10^3 \times 0.5 \cdot 10^{-6} = 1$ ms

When the transistor T_4 is ON, C_4 will discharge through it and the equivalent resistance: $R_{equiv} = 1k\Omega \parallel 3\ k\Omega \parallel 1.8\ k\Omega \parallel 39\ k\Omega \parallel h_{11eT5} = 0.5\ k\Omega$

with the time constant: $\tau = R_{equiv} \times C_4 = 0.5 \cdot 10^3 \times 0.5 \cdot 10^{-6} = 0.25$ ms

The positive impulse appears on the T_5 base and turn T_5 OFF.

When T_5 is ON the resistor divider from the T_5 base arranges a voltage $U_{BET5} = 0.88$ V and consequently a current $I_{CT5} = 14.7$ mA.

Considering that the resistance of the transformer primary Tr_2 is neglected, then $U_{CE} = 20$ V.

So, in the time of conducting, the power dissipated on T_5 is

P = 20 V x 15 mA = 300 mW,

an acceptable power for a medium power transistor as BD140.

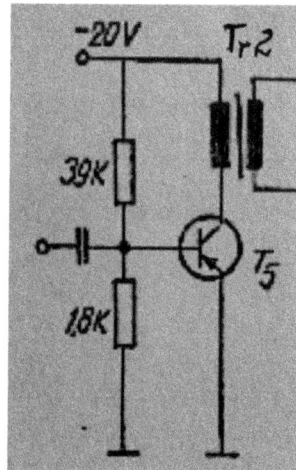

Fig 10

The output transformer Tr_2, see fig 10, is a low power transformer because in its secondary it is necessary a current 40 mA and a voltage 1 V

to open the thyristor T: $P_{Trsec} = 1 \times 40 \times 10^{-3} = 40$ mW

It is chosen the transformer core strip E8 and the core transversal cutting, cross section 1.6 x 0.8 cm^2.

Will result the number of winds per volt, considering the repetitive process with the frequency 100 Hz:

$n' = 10^8 / (4.44 \cdot B_{max} \cdot S_m \cdot f)$

$n' = 10^8 / (4.44 \times 1.1 \times 10^4 \times 1.6 \times 0.8 \times 100) = 16$

$n'' = 1.1 \times 16 = 17.6$

$U_1 = 20$ V $U_2 = 1$ V

$n_1 = 16 \times 20 = 320$ winds $n_2 = 18 \times 1 = 18$ winds

$\varnothing_1 = 0.1$ mm $\varnothing = 0.2$ mm

Diodes Bridge $D_5 D_6 D_7 D_8$

The bridge has to transform the alternating voltage of the powerline in the positive voltage to turn ON the thyristor T.

The diodes are type 6514.

The thyristor is KT 705.

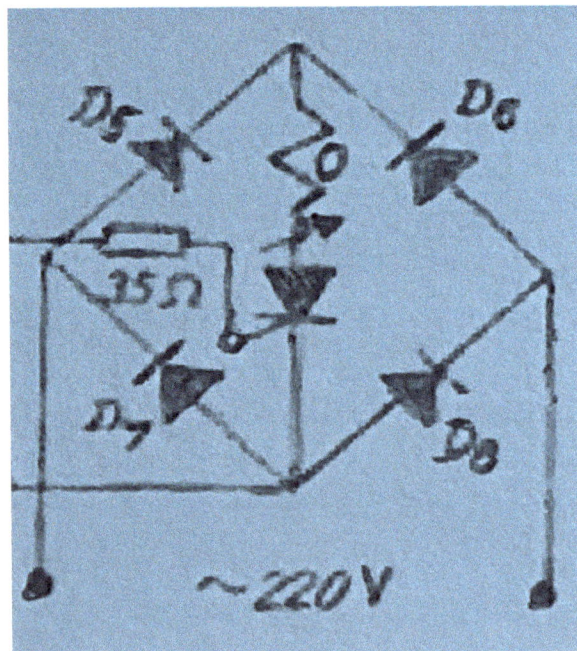

Fig 11

Operational Amplifier A

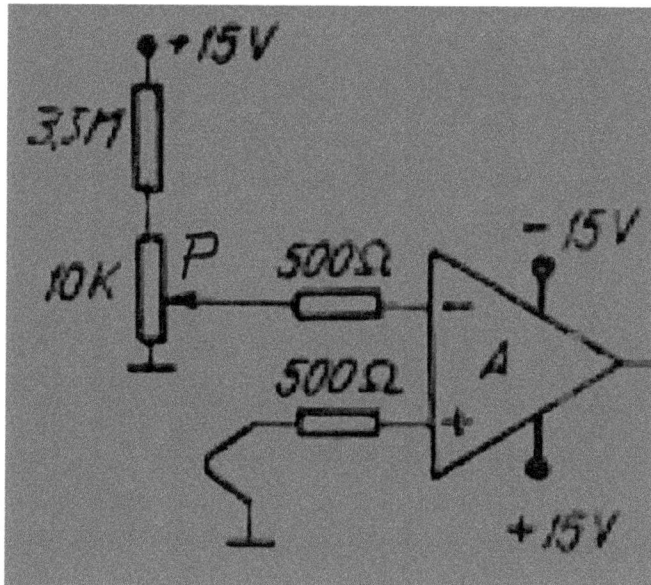

Fig 12

The operational amplifier A has to compare the voltage given by the thermocouple and the voltage given by the helical potentiometer, helipot P. Their difference is amplified by A and activates the opening system of the thyristor in such a way that the temperature increases/decreases to minimise this difference; it is a negative feedback system.

A is type μA 741.

Thermocouple Chromel-Alumel has the following temperature characteristics:

TABLE

0° C	0.00 mV	300° C	12.21 mV
100° C	4.10 mV	400° C	16.40 mV
200° C	8.13 mV	500° C	20.65 mV

The helipot P is so designed that the voltage given by the position of its wiper corresponds to a certain temperature.

The helipot P of 10 kΩ is supplied at +15 V through a resistance of 3.5 MΩ. By design, it is possible to arrange the oven temperature by the position of the wiper of the helipot P:

On the position 1. of the helipot P, the oven will be heated at 100° C

On the position 2. of the helipot P, the oven will be heated at 200° C

On the position 3. of the helipot P, the oven will be heated at 300° C
....

For intermediate positions to those already presented will be obtained inter-temperatures to those given, the temperature variation being practically linear.

Heating Technique of Oven

The thyristor is supplied by positive semi waves powerline voltage obtained by rectifying work of the diodes $D_5 D_6 D_7 D_8$.

When the thyristor is open by positive impulses on its gate, its current I which supplies the heating resistor of the oven, looks as in fig 11 with maximum I_{max}. The medium electric current I_{med} which heats the oven is function of angle θ:

$$I_{med} = 1/\pi \int_{\pi-\theta}^{\pi} I_{max} \sin (wt)\, dwt = (-1/\pi) I_{max} \cos (wt) \Big|_{\pi-\theta}^{\pi} =$$

$$= (-1/\pi) I_{max} (-1 - \cos (\pi - \theta)) = (1/\pi) I_{max} (1 - \cos \theta)$$

$$\partial I_{med}/\partial \theta = (I_{max}/\pi) \sin \theta$$

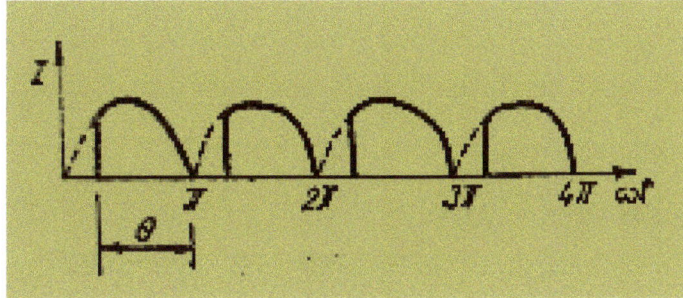

Fig 11

TABLE

$\Theta = \pi$	$\cos \theta = -1$	$1-\cos \theta = 2$	$I_{med} = (2/\pi)\, I_{max}$
$\Theta = 3\pi/4$	$\cos \theta = -\sqrt{2}/2$	$1-\cos \theta = 1.7$	$I_{med} = (1.7/\pi)\, I_{max}$
$\Theta = \pi/2$	$\cos \theta = 0$	$1-\cos \theta = 1$	$I_{med} = (1/\pi)\, I_{max}$
$\Theta = \pi/4$	$\cos \theta = \sqrt{2}/2$	$1-\cos \theta = 0.3$	$I_{med} = (0.3/\pi)\, I_{max}$
$\Theta = 0$	$\cos \theta = 1$	$1-\cos \theta = 0$	$I_{med} = 0$

Considering that the variation of angle θ in the functioning time is around $\pi/2$. the variation of the medium current follows the variation of angle θ:

for $\theta = \pi/2$
$\sin \theta = \sin \pi/2 = 1$
$\cos \theta = 0$
$1 - \cos \theta = 1 - \cos \pi/2 = 1$

$$I_{med} = (1/\pi)\, I_{max}\, (1 - \cos \theta)$$

$$\partial I_{med}/\partial \theta = (I_{max}/\pi)\, \sin \theta$$

$$\Delta I_{med} = (I_{max}/\pi)\, \Delta\theta$$

$$\Delta\theta = \pi\, \Delta I_{med}/I_{max} = \Delta I_{med}/I_{med}$$

$$\Delta I_{med} = \Delta\theta\, I_{med}$$

97

The equality between the voltage fixed by the helicoidal potentiometer P and the voltage given by the thermocouple is accomplished only for a short time, because in the moment of equality the A output is null and will cease the command to heat the oven.

The temperature is decreasing, will appear an inequality of voltages at input which amplified gives at operational amplifier output a negative voltage, which commands again the heating.

Since the big amplification of the operational amplifier A ($20000 \div 100000$) are sensed and corrected fast the deviations from the chosen temperature. Practically, because the inertia and thermal losses of the oven, the deviation from the chosen temperature was observed to be $\sim 1\ \%$ from the work temperature.

The thermoregulator is built function of the desired oven, that decides electrical currents, voltages, power etc. for the thyristor T and the diodes bridge D5 D6 D7 D8.

references

Translated Article: TERMOREGULATOR CU TERMOCUPLU by Irina Rabeja
                    Journal Electrotehnica Electronica Automatica BUCURESTI ANUL 29 Nr. 3 APRILIE 1981
Wikipedia

THERMOREGULATOR FOR THE FILAMENTS IN VACUUM

The early observers of the natural phenomena differentiated qualitatively the hot and cold objects, on the basis that hot objects give heat to cold objects. Matter from the heating degree view point is characterised by a state called temperature.

The temperature recording of a system can be done function of the physical properties of a reference standard object brought at thermal equilibrium with the system.

Properties as physical state, thermal expansion, electrical resistance, crystal shape are the base for many temperature scales.

Measuring, recording and regulation of temperature in research and industry systems are done nowadays with high precision by using electronic devices.

Electronic temperature regulators or thermoregulators are used in many places, like electric ovens, thermostats or filaments in vacuum.

The last application is useful:

- o to study of surface interactions at atomic scale
- o to elucidate the reaction mechanisms gas-metal in heterogenous catalysis
- o to elucidate the formation mechanisms of epitaxial crystal layers for depositing semiconductor elements
- o to study the corrosive action of gases on different materials (metals)

Part of those studies can be done with the *Microscope for Field Emission*. That is a small installation realised with a glass cell at the low pressure of 10^{-9} torr having inside a cathode filament made of metal wolfram, symbol W, known also as tungsten.

The cathode filament is a heated wire that acts as a source of electrons.

The filament is at 4000V electric potential difference from an anode formed by a fluorescent disk. In this electric field appears an electronic current of 10^{-12} A which gives an image on the anode surface, representing the distribution of atoms in the crystal net (structure) surface of wolfram metal.

In the same cell there is a filament made of metal silver, symbol Ag, which, when heated, emits silver atoms which can deposit on the wolfram wire.

Those atoms can migrate on the wolfram surface function of its temperature. In this way it is studied the adsorption, function of time and temperature, of a metal on other metal, the migration of the adsorption layer on the wolfram surface as function of time and temperature, the desorption and the controlled evaporation of silver or other metal in void - the density of atoms/cm^2 in unit of time.

And, for ex. in the case of adsorption of metals on metals, like silver on wolfram, will be obtained results as:

- o The mechanism of forming and growing of Ag layers on W
- o The influence of atomic structure of W layer on structure of Ag layers
- o The connection energy Ag-W

Alike it is done the study of desorption of different types of gases, adsorbed on wolfram.

For that it is used a glass balloon with vacuum, in which there is a wolfram filament heated by an electrical current.

It is introduced the gas under study at the pressure 10^{-9} torr.

The gas is adsorbed on the wolfram wire surface at the room temperature.

Then the wolfram filament is heated at the temperature T. At that temperature are desorbed from the surface a certain type of ions or atoms. Those ions can be registered (function of their atomic masses) in the mass spectrometer.

In this way it is possible to show the desorption energy of different types of gases adsorbed on the studied surface.

The wolfram filament is heated by the current generated by a thermoregulator - temperature regulator - and in the same time its temperature is controlled, using as sensor for temperature its electrical resistance.

The electrical resistance R of the filament is deduced by measuring the voltage V on it and also measuring the current I through it: $R = V / I$

In the specialty literature for wolfram/tungsten W, there are diagrams with precise curves for its electrical resistance dependency of temperature:

$R_T / R_0 = f(T)$ - see fig 1

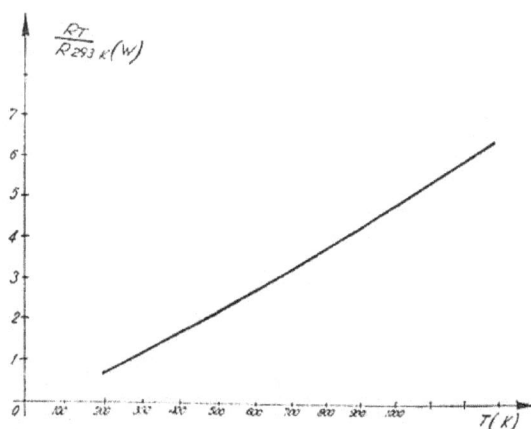

Fig 1

R_0 is the electrical resistance at reference temperature 20°C or 293 K

R_T is the electrical resistance at temperature T

T is the temperature in Kelvin degrees

The relationship between the electrical resistance of tungsten and the temperature can be described by the formula $R=R_0 (1+a\Delta T)$ where R_0 is the resistance at the reference temperature, a is the temperature coefficient of resistance for tungsten $\sim 4.5x10^{-3}/ °C$ and ΔT is the change of temperature. This formula indicates that as the temperature of tungsten increases, its electrical resistance also increases.

An automatic control of temperature can be done with the thermoregulator with the schematic diagram presented in fig 3, comparing the voltage between two points of the wolfram filament with that generated in a known load and amplifying the signal difference.

It is considering that the ADCB part of the wolfram filament in fig 2 is circulated by the electric heating current, which circulates in the same time in the potentiometer P.

A potentiometer P is a manually adjustable variable resistor with 3 terminals. Two of the terminals are connected to the opposite ends of a resistive element with electric resistance $R_{Ptot.}$ and the third terminal connects to
a sliding contact, called a wiper, moving over the resistive element, with electric resistance R_P between wiper and one of the terminals.

$R_{Ptot.}$ is the resistance between the two terminals of potentiometer P

R_P is the resistance between the wiper and one of the terminals of P

U_{RPtot} is the voltage on the terminals of potentiometer P

U_{RP} is the voltage between the wiper and one of the terminals of P

The resistances which are compared are the DC part of the filament and the resistance R_P.

The voltage on DC part is introduced in a compensation system, so when $U_{DC} = U_{Rp}$ the voltages applied to the inputs of an operational amplifier, the voltage U_I and the voltage U_{II}, should be equal.

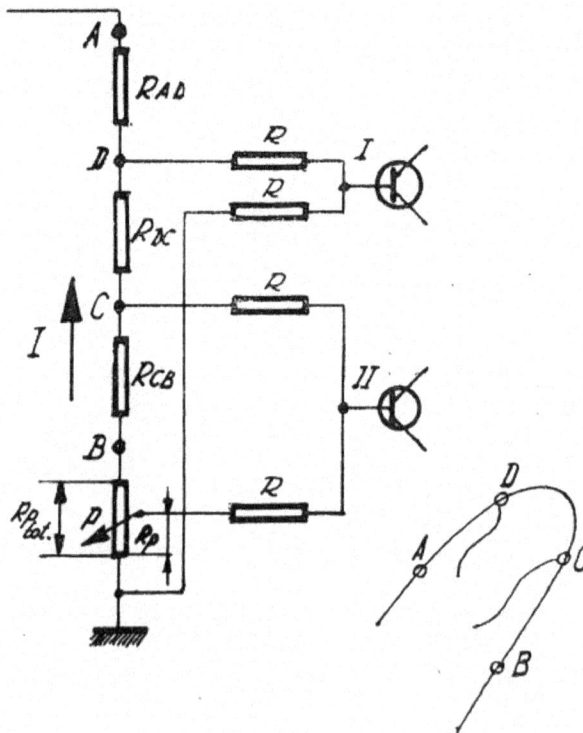

Fig 2

In the case $U_I \neq U_{II}$ the amplifier which is in a loop with negative reaction with the power heating system of the filament, amplifies the potential difference provoking - with further amplification and five power emitter repeaters in parallel - the modification of the current in filament until the two voltages U_I and U_{II} become equal:

$$U_I = R(U_{DC} + U_{CB} + U_{RPtot})/2R = (U_{DC} + U_{RPtot} + U_{CB})/2$$

$$U_{II} = R(U_{CB} + U_{RPtot} - U_{RP})/2R + U_{RP} = (U_{CB} + U_{RPtot} + U_{RP})/2$$

$$U_I = U_{II} \quad \text{when } U_{DC} = U_{RP}$$

$$U_I = (IR_{DC} + IR_{CB} + IR_{Ptot})/2 = I(R_{DC} + R_{CB} + R_{Ptot})/2$$

$$U_{II} = (IR_{CB} + IR_{Ptot} + IR_P)/2 = I(R_{CB} + R_{Ptot.} + R_P)/2$$

$$U_I = U_{II} \quad \text{when } R_{DC} = R_P$$

R_{DC} is changing with its temperature.

The complete electronic schematic diagram of this thermoregulator is presented in fig 3.

Initial a current is arranged with the variable resistor, potentiometer P_1 seen in fig 3, with the device on position Manual.

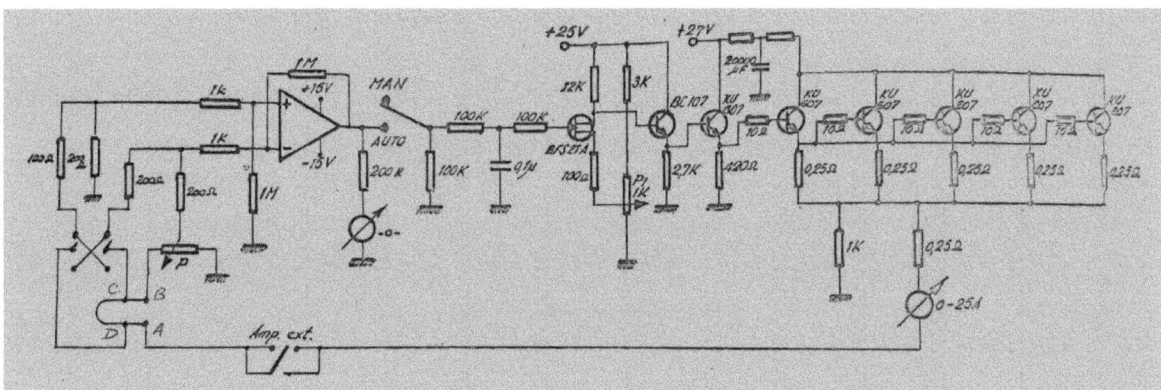

Fig 3

The potentiometer P is positioned at a certain value R_P and since the above-described functioning, the electrical resistance of the DC part of the filament will have in short time the same value, meaning that the wolfram filament is brought at a temperature which will be maintained constant.

So the potentiometer P range can be calibrated directly in temperature values. In functioning, when the potentiometer P is positioned at the desired temperature, the wolfram filament is brought at the desired temperature which will be maintained constant.

The advantages of the above electronic control are:

1. Electrical current in the filament can be regulated between 0-25 A

2. The regulated temperature precision: error < 3 %

   To calculate the stability of the temperature it is necessary to know the value of the R_{DC} resistance, so it is measured U_{DC} and $I_{filament}$ and it is done their ratio:

   $$R_{DC} = U_{DC}/I_{filament} = f(T) \qquad R_{DC} = R_T \qquad R_T = U_{DC}/I_{filament} = f(T)$$

   From the curve $R_T/R_0 = f(T)$ is deduced the temperature T of the filament. Using the device on the position AUTO are obtained the following values:

For	U_{DC}	318	320	323	325	331	mV
	I	2.152	2.163	2.189	2.202	2.231	A

Results:	R_T	147.77	147.94	147.55	147.59	148.36	mΩ

   average $R_T = 147.84$ mΩ

   It is observed $\Delta R_T < 1$ mΩ

   $R_0 = 31$ mΩ

   $R_T/R_0 = 147.84/31 = 4.76$ –> reading the diagram for electrical resistance dependency of temperature for wolfram $R_T / R_0 = f(T)$, resulted T = 1035 K

   $\Delta R_T < 1$ mΩ

   $\Delta T = \Delta R_T/R_0$

   $\Delta R_T/R_0 < 1/31 = 0.03$ –> $\Delta T < 3$ %

3. Response time:

On AUTO position, at the automatic modification of the work temperature point, the time to reach the desired temperature or the response time is very small, 5–20 s.

On MANUAL position, the time to reach the desired temperature or the response time could reach 1.5 min.

4. Reproducibility:

Always it is the same temperature at the same position of variable resistor R_P.

references

Translated Article: APARAT PENTRU REGLAREA AUTOMATA A TEMPERATURII FILAMENTELOR IN VID by Irina Rabeja
Journal Electrotehnica Electronica Automatica BUCURESTI ANUL 26 Nr. 6 AUGUST 1978
Wikipedia

ANNEX

File:Bio-inspired Big Dog quadruped robot is being developed as a mule that can traverse difficult terrain.tiff

From Wikimedia Commons, the free media repository

Summary

Description	**English:** Bio-inspired Big Dog quadruped robot is being developed as a mule that can traverse difficult terrain
Date	23 August 2012
Source	This file was derived from: DARPA Strategic Plan (2007).pdf
Author	DARPA

Licensing

*This image or file is a work of a Defense Advanced Research Projects Agency (DARPA), an agency of the United States Department of Defense, employee, taken or made as part of that person's official duties. As a work of the U.S. federal government, the image is in the **public domain**.*

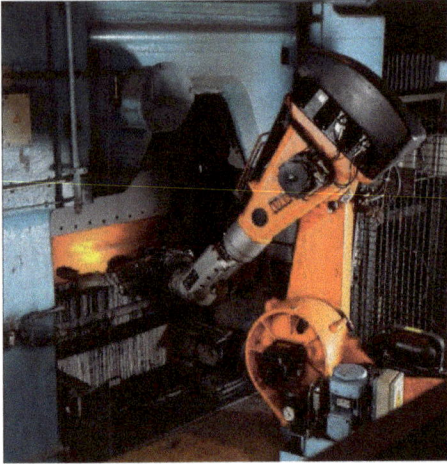

File: Automation of foundry with robot.jpg

From Wikimedia Commons, the free media repository

Summary

Description	Factory Automation with industrial robots for metal die casting in foundry industry, robotics in metal manufacturing
Date	2003
Source	KUKA Roboter GmbH, Zugspitzstraße 140, D-86165 Augsburg, Germany, Dep. Marketing, Mr. Andreas Bauer, http://www.kuka-robotics.com
Author	KUKA Roboter GmbH, Bachmann

Licensing

File:IED detonator.jpg
From Wikimedia Commons, the free media repository

Description	**English:** IED DETONATOR — A U.S. Marine Corps explosive ordnance disposal technician prepares to deploy a device that will detonate a buried improvised explosive device near Camp Fallujah, Iraq, Nov. 27, 2005. The Marine is assigned to Combat Logistics Brigade 8, 2nd Marine Logistics Group. U. S. Marine Corps photo by Lance Cpl. Bobby J. Segovia.
Date	27 November 2005
Source	U.S. Dept. of Defense here.
Author	Lance Cpl. Bobby J. Segovia
Permission (Reusing this file)	*This file is a work of a United States Marine or employee, taken or made as part of that person's official duties. As a work of the U.S. federal government, it is in the **public domain**.* العربية · বাংলা · català · Deutsch · English · español · français · magyar · italiano · 日本語 · македонски · മലയാളം · မြန်မာဘာသာ · Nederlands · português · русский · sicilianu · slovenščina · svenska · Türkçe · українська · Tiếng Việt · 中文（简体） · 中文（繁體） · +/−

File:Factory Automation Robotics Palettizing Bread.jpg

KUKA industrial robots being used at a bakery for food production

Size of this preview: 799 × 599 pixels. Other resolutions: 320 × 240 pixels | 640 × 480 pixels | 1,024 × 768 pixels | 1,280 × 960 pixels | 2,560 × 1,920 pixels | 3,417 × 2,563 pixels

Licensing

[edit]

This work has been released into the **public domain** by its author, **KUKA Roboter Gmbh**. This applies worldwide.

In some countries this may not be legally possible; if so:

*KUKA Roboter Gmbh grants anyone the right to use this work **for any purpose**, without any conditions, unless such conditions are required by law.*

File history

Click on a date/time to view the file as it appeared at that time.

	Date/Time	Thumbnail	Dimensions	User	Comment
current	**06:14, 25 May 2021**		3,417 × 2,563 (1.85 MB)	Dekema (talk | contribs)	discovered red higher resolution copy
	20:32, 23 December 2017		1,024 × 768 (473 KB)	Dekema (talk | contribs)	much higher resolution

File:Alan Turing (1951).jpg

Size of this preview: 449 × 599 pixels. Other resolutions: 180 × 240 pixels | 360 × 480 pixels. Original file (800 × 1,067 pixels, file size: 91 KB, MIME type: image/jpeg)Open in Media Viewer

File information

Licensing

This work is in the public domain in its **source country** for the following reason:
This work is in the public domain in the **United States** for the following reason:

*This non-U.S. work was published in 1930 or later, but is in the public domain in the **United States** because it was simultaneously published (with 30 days) in the U.S. and in its source country and is in the public domain in the U.S. as a U.S. work (no copyright registered, or not renewed).*

For background information, see the explanations on Non-U.S. copyrights. **Note:** in addition to this statement, there *must* be a statement on this page explaining *why* the work is in the public domain in the U.S. Additionally, there must be verifiable information about previous publications of the work.

his work in the public domain in my country?

rent	08:10, 5 September 2024		800 × 1,067 (91 KB)	The way of Changpian (talk | contribs)

File:Francesco Melzi - Portrait of Leonardo.png

Licensing

This is a faithful photographic reproduction of a two-dimensional, public domain work of art. The work of domain for the following reason:

This work is in the public domain in its country of origin and other countries and areas where author's life plus 100 years or fewer.

This work is in the public domain in the United States because it was published (or registered Office) before January 1, 1930.

This file has been identified as being free of known restrictions under copyright law, including all related a

The official position taken by the Wikimedia Foundation is that *"faithful reproductions of two-dimensional public domain"*.

This photographic reproduction is therefore also considered to be in the public domain in the United States use of this content may be restricted; see Reuse of PD-Art photographs for details.

File history

Click on a date/time to view the file as it appeared at that time.

	Date/Time	Thumbnail	Dimensions	User	Comment
current	01:06, 8 May 2019		1,286 × 1,842 (4.25 MB)	UpdateNerd (talk | contribs)	more hi-res version

File: Vulcan by Guillaume Coustou the Younger Louvre MR1814.jpg

Vulcan/Hephaestus by Guillaume Coustou the Younger

Summary

Artist	**Guillaume Coustou the Younger** (1716–1777) ✏ ▮▮▮  **Français :** Guillaume II Coustou (Français, 1716-1777)
Description	**English:** *Vulcan*. Marble, reception piece for the French Royal Academy, 1742. **Français :** *Vulcain*. Marbre, pièce de réception à l'Académie royale, 1742.
Dimensions	H. 69 cm (27 in.), W. 50 cm (19 ½ in.), D. 41 cm (16 in.)
Collection	**Louvre Museum** ✏ ▮▮▮ (**Inventory**)
Current location	Department of Sculptures, Richelieu, ground floor, room 25
Accession number	MR 1814
Credit line	Seized during the French Revolution
Source/Photographer	Jastrow (2006)

Licensing

File:PIA15279 3rovers-stand D2011 1215 D521.jpg

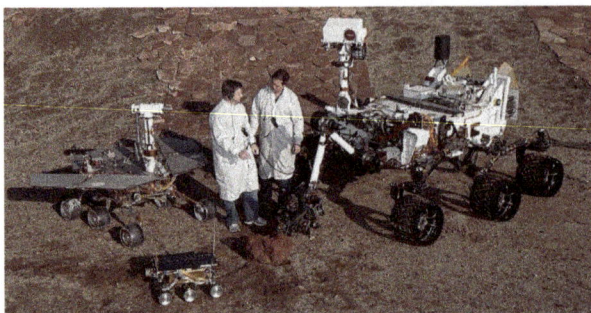

Size of this preview: 800 × 404 pixels. Other resolutions: 320 × 162 pixels | 640 × 323 pixels | 1,024 × 517 pixels | 1,280 × 646 pixels | 2,560 × 1,292 pixels | 5,723 × 2,889 pixels. Original file (5,723 × 2,889 pixels, file size: 7.34 MB, MIME type: image/jpeg)Open in Media Viewer

Licensing

*This file is in the **public domain** in the United States because it was solely created by NASA. NASA copyright policy states that "NASA material is not protected by copyright **unless noted**". (See Template:PD-USGov, NASA copyright policy page or JPL Image Use Policy.)*

Warnings:
Use of NASA logos, insignia and emblems is restricted per U.S. law 14 CFR 1221.
The NASA website hosts a large number of images from the Soviet/Russian space agency, and other non-American space agencies. These are *not necessarily* in the public domain.
Materials based on Hubble Space Telescope data may be copyrighted if they are not explicitly produced by the STScI.[1] See also {{PD-Hubble}} and {{Cc-Hubble}}.
The SOHO (ESA & NASA) joint project implies that all materials created by its probe are copyrighted and require permission for commercial non-educational use. [2]
Images featured on the *Astronomy Picture of the Day* (APOD) web site may be copyrighted. [3]
The National Space Science Data Center (NSSDC) site has been known to host copyrighted content. Its photo gallery FAQ states that all of the images in the photo gallery are in the public domain "Unless otherwise noted."

File history
Click on a date/time to view the file as it appeared at that time.

	Date/Time	Thumbnail	Dimensions	User	Comment
current	11:02, 25 January 2012		5,723 × 2,889 (7.34 MB)	Nova1 3	cropped

File:HST-SM4.jpeg

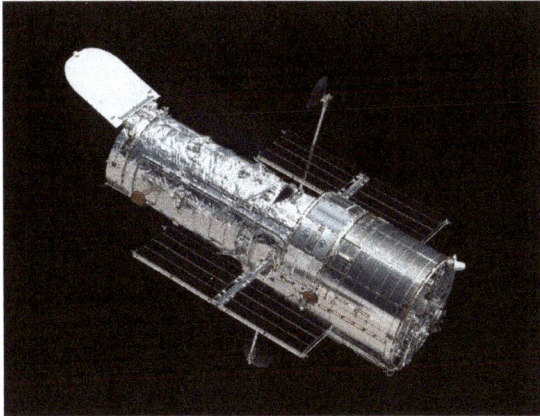

Original file (2,022 × 1,518 pixels, file size: 318 KB, MIME type: image/jpeg)
Summary
Licensing

*This file is in the **public domain** in the United States because it was solely created by NASA. NASA copyright policy states that "NASA material is not protected by copyright **unless noted**".*
(See Template:PD-USGov, NASA copyright policy page or JPL Image Use Policy.)

File:Leonardo-Robot3.jpg

Model of Leonardo's robot with inner workings.
Possibly constructed by Leonardo da Vinci around 1495.

Captions

English

Add a one-line explanation of what this file represents

Description	Model of a robot based on drawings by Leonardo da Vinci.
Date	Berlin 2005 (30 October 2005 (from Exif))
Source	Own work
Author	Photo by Erik Möller. *Leonardo da Vinci. Mensch - Erfinder - Genie* exhibit, B

File history

	Date/Time	Thumbnail	Dimensions	User	Comment
current	14:21, 4 June 2007		1,370 × 1,370 (272 KB)	Rama (talk | contribs)	

File:HEINRICH HERTZ.JPG

Size of this preview: 414 × 479 pixels.

File:Albert Einstein (Nobel).png

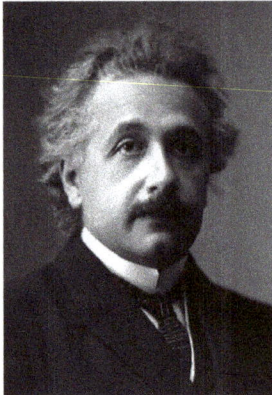

[Albert Einstein (Nobel).png](#) (280 × 396 pixels, file size: 94 KB, MIME type: image/png)

This is a file from the Wikimedia Commons. Information from its **description page the**
Commons is a freely licensed media file repository. You can help.

Description	**English:** Albert Einstein, official 1921 Nobel Prize in Physics photograph. **Français :** Albert Einstein, photographie officielle du Prix Nobel de Physique 1921.
Date	1921
Source	Official 1921 Nobel Prize in Physics photograph
Author	Unknown
Permission (Reusing this file)	*This media file is in the **public domain** in the United States. This applies to U.S. works where the copyright has expired, often because its first publication occurred prior to January 1, 1923. See this page for further explanation.* This image might not be in the public domain outside of the United States; this especially applies in the countries and areas that do not apply the rule of the shorter term for US works, such as Canada, Mainland China (not Hong Kong or Macao), Germany, Mexico, and Switzerland. The creator and year of publication are essential information and must be provided. See Wikipedia:Public domain and Wikipedia:Copyrights for more details. English	español	suomi	italiano	македонски	português do Brasil	română	py

File:Crystal oscillators

Symbol of Quartz Crystal
Electrical Equivalent of Quartz Crystal

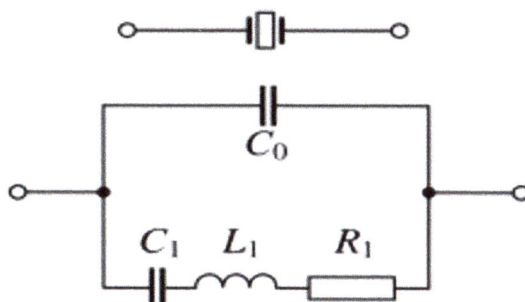

Licensing

ͽ:enago academy **15** YEARS OF TRUST
Learn. Share. Discuss. Publish

" Is Translating an Already Published Article Ethical?

Yes, it is – if you declare that it is a translated version of an already published paper."

At Enago Academy, our mission is to empower researchers worldwide by equipping them with the knowledge and skills necessary for successful academic writing, publishing, and career advancement. We are passionate about helping early-career researchers overcome challenges in publishing their work in high impact journals, as well as supporting early-stage and experienced researchers in building their research profiles and maximizing opportunities for overall growth.

We believe that there is a better way to aid academics in fulfilling their research goals through a more practical, multilingual, up-to-date, and less intrusive avenue where they can gain insightful knowledge and acquire skills to achieve publication success.

AI overview: It is a normal part of the design process as prototypes are for testing ideas, can be done in one location or country while final testing, design iteration or publication can occur elsewhere. This is a common and effective practice for designers and businesses to work internationally, leveraging different resources or expertise in different regions.